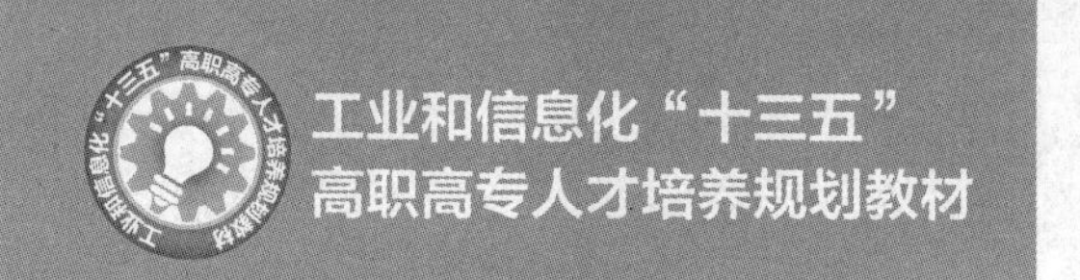

计算机专业英语 第2版

Computer Professional English

朱龙 刘长君 ◎ 主编
孙雅妮 谢宇 ◎ 副主编
赵克林 ◎ 主审

人民邮电出版社
北京

图书在版编目（CIP）数据

计算机专业英语 / 朱龙，刘长君主编. -- 2版. -- 北京 : 人民邮电出版社，2018.8（2021.3重印）
工业和信息化“十三五”高职高专人才培养规划教材
ISBN 978-7-115-48069-9

Ⅰ. ①计… Ⅱ. ①朱… ②刘… Ⅲ. ①电子计算机－英语－高等职业教育－教材 Ⅳ. ①TP3

中国版本图书馆CIP数据核字(2018)第047978号

内容提要

本书共 7 章，涵盖了英语在计算机硬件、软件、网络等多方面的应用，主要包括硬件基础、计算机系统维护、计算机网络基础、软件、程序设计语言、计算机网络技术、IT 职场英语等，图文并茂，中英对照，生动易读。书中还加入课文朗读音频二维码，读者扫码即可开展移动学习。

本书可作为高职高专院校计算机相关专业计算机专业英语课程教材，也适合职业培训机构和自学者使用。

◆ 主　　编　朱　龙　刘长君
副 主 编　孙雅妮　谢　宇
主　　审　赵克林
责任编辑　桑　珊
责任印制　马振武

◆ 人民邮电出版社出版发行　　北京市丰台区成寿寺路 11 号
邮编　100164　　电子邮件　315@ptpress.com.cn
网址　http://www.ptpress.com.cn
三河市君旺印务有限公司印刷

◆ 开本：787×1092　1/16
印张：11.75　　2018 年 8 月第 2 版
字数：268 千字　　2021 年 3 月河北第10次印刷

定价：35.00 元

第2版前言 FOREWORD

通过在高等职业院校多年计算机专业课的教学，我们体会到：英语是专业课学习的难点和障碍之一。为此，我们编写了这本计算机专业英语教材，编者们均是长期从事计算机专业课教学的教师和资深英语教师，在内容上进行了大刀阔斧的改革，以区别于同类教材。本书出版5年来，已经重印11次，受到许多职业院校师生的欢迎。此次修订，在保留原书特色的基础上，对存在的一些问题加以修正，对部分章节进行了完善和更新，加入了课文朗读音频二维码，方便读者移动学习。

本书有如下特色：

1. 本书讲解了丰富的计算机知识，并且非常实用、全面，旨在为读者学习后续课程（如计算机语言、硬件与网络维护、计算机网络应用、图形图像处理等）扫清障碍。

2. 版面活泼，图文并茂，布局简洁流畅，赏心悦目，具有很好的视觉效果。

3. 分散难点，突出重点，在各节（页）后罗列新出现的专业词汇，绝不堆砌大量的专业词汇，避免学生产生畏难情绪。

4. 精编与计算机新技术结合的短文、英汉对照阅读，每章后讲解若干专业术语，以方便任课教师往纵深发挥。

5. 各章安排相关的情景对话，使学生易学易懂，并且随处都能开口讲，从而调动学生的学习兴趣。

6. 各章内容难度依次加深，适合各种层次的学生，基础好的学生可以阅读章后附文，以了解更多的专业词汇。

7. 作业形式多样，难度适中。

本书在内容和选择方面，特别考虑了基础性与专业性的结合，因此在教学中，建议将第1、2、3、7章作为基础性的必学内容，而第4、5、6章可根据不同专业需要作为选学内容。

本书由朱龙、刘长君担任主编，孙雅妮、谢宇担任副主编，赵克林担任主审，向文欣、乔治锡、肖朵参与了本书的编写。本书录音由黄秀英、程元元录制。

由于水平有限，书中难免有疏漏或不足之处，敬请读者批评指正。

编者

2017年11月

目 录 CONTENTS

Chapter 1 Hardware Concepts（硬件基础）

教学要求

掌握专业关键词汇（key words）；能阅读本章所列英语短文；能识别计算机的各组件。

教学内容

硬件英语；各相关硬件设备品牌、主要生产厂商；常用的专业术语。

教学提示

到学校机房或本地电脑城参观，感受本章内容，以学到更多的专业词汇。

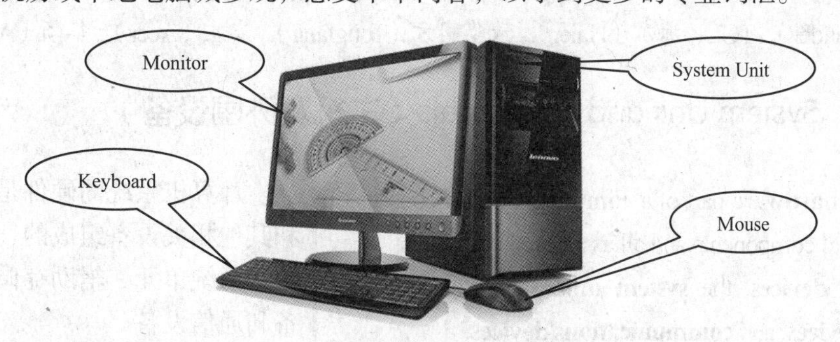

1.1 Personal Computer（个人计算机）

A **personal computer**, or PC, is a type of **microcomputer** designed for the use by only one person at a time. Generally, a PC is a small **digital computer** constructed around a **microprocessor**, which is a semiconductor **chip** that contains all the arithmetic, logic, and control circuitry needed to perform the functions of a computer's **central processing unit** (CPU).

个人计算机简称为 PC，是一种专为个人用户设计的微型计算机，一次只能由一个人使用。一般来说，PC 是一种以微处理器为中心构建起来的小型数字计算机。微处理器本身是一块半导体芯片，包含各种算术、逻辑、控制电路，用以实现中央处理器（CPU）的各项功能。

By 1990 some personal computers had become small enough to be completely portable; these included **laptop computers,** which could rest on one's lap; **notebook computers**, which were about the size of a notebook; and pocket or palm-sized computers, which could be held in one's hand.

到 1990 年，某些个人计算机已是十分小巧了，完全可以随身携带，如可以放置在膝上使用的便携式计算机，笔记本大小的笔记本电脑，还有可以握在手中使用的袖珍型（或称为掌上型）计算机。

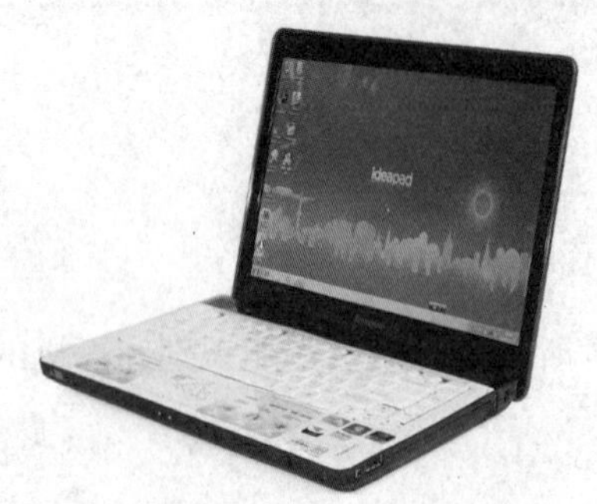

Notebook Computer

 **Key words:** personal computer（个人计算机），microcomputer（微型计算机），digital computer（数字计算机），microprocessor（微处理器），chip（芯片），central processing unit（中央处理器），laptop computer（便携式计算机），notebook computer（笔记本电脑）

常见的计算机品牌：苹果（Apple）、联想（Lenovo-IBM）、戴尔（DELL）、惠普（HP）、方正（Founder）、TCL、海尔（Haier）、清华同方（Tongfang）、宏碁（Acer）、华硕（ASUS）等。

1.2 System Unit and Peripherals（主机及外围设备）

The **hardware** part of a **computer system** consists of the physical components and all associated equipment. They are **input devices**, the **system unit**, **secondary storage**, **output devices**, and **communications devices**.

计算机系统的硬件是由物理部件和其他相关设备组成的，包括输入设备、系统单元、辅助存储器、输出设备和通信设备。

Input devices are equipment that translates **data** and **programs** that humans can understand into a form that the computer can process. The most common input devices are the **keyboard**, **scanner**, **input pen**, **touch screen** and **mouse**. The keyboard on a computer looks like a typewriter keyboard, but it has additional specialized keys. A mouse is a device that typically rolls on the desktop. It directs the insertion point, or **cursor**, on the display screen. A mouse has one or more buttons for selection commands. It is also used to draw figures.

输入设备的功能是将人所能识别的数据和程序翻译成计算机能够处理的形式。最常用的计算机输入设备包括键盘、扫描仪、输入笔、触摸屏和鼠标。计算机上的键盘就像打字机上的键盘一样，只不过另外有一些专用键位。鼠标是在桌面上移动的一种装置，用以在显示屏上控制插入点或移动光标。鼠标有一个或多个按键，用于选择命令。鼠标还可以用来绘制图形。

① Monitor
② Keyboard
③ Mouse
④ CD-ROM
⑤ Scanner
⑥ Printer
⑦ Speaker
⑧ Disk
⑨ Megaphone
⑩ Camera
⑪ USB drive

朗读音频

System Unit and Peripherals

 **Key words:** hardware（硬件），computer system（计算机系统），input device（输入设备），system unit（系统单元），secondary storage（辅助存储器），output device（输出设备），communications device（通信设备），data（数据），program（程序），keyboard（键盘），scanner（扫描仪），input pen（输入笔），touch screen（触摸屏），mouse（鼠标），cursor（光标）

1.2.1 Monitor/Display（显示器）

A **display** is a computer output **surface** that shows text and often graphic **images** to the computer user, using a **cathode ray tube (CRT), liquid crystal display (LCD)** or other image projection technology. A display can be distinguished according to: **color** capability, **sharpness** and viewability, screen size, and the **projection technology**.

显示器是一种计算机输出界面，用以将文本和图像呈现给计算机用户。显示器所使用的图像显像技术包括阴极射线管（CRT）技术、液晶显示（LCD）技术等。显示器主要有以下几项技术指标：色彩指标、清晰度、可视性、屏幕尺寸以及显像技术。

 **Key words:** monitor（显示器），display（显示器），surface（界面），image（图像），CRT（阴极射线管），LCD（液晶显示器），color（色彩）；sharpness（清晰度），projection technology（显像技术）

Monitor (LCD)

Monitor (CRT)

朗读音频

Monitor

 常见的显示器品牌：三星（SAMSUNG）、AOC、戴尔（DELL）、飞利浦（Philips）、LG、HKC、惠普（HP）、宏碁（Acer）、优派（ViewSonic）、明基（BenQ）、华硕（ASUS）、三菱、长城（Great Wall）、Eizo、NEC、苹果（Apple）等。

1.2.2 Input Device: Keyboard and Mouse（输入设备：键盘和鼠标）

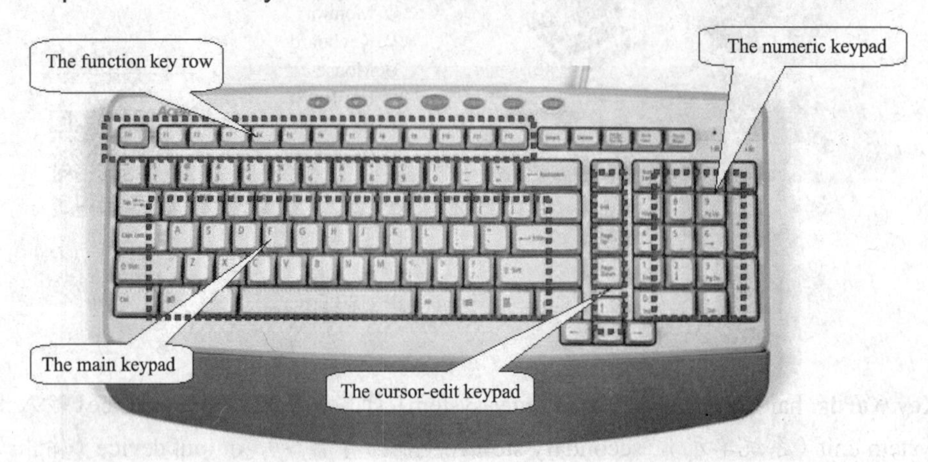

The Keyboard

On most computers, the keyboard is the primary text input device. (The mouse is also a primary input device but its ability to easily transmit textual information is limited.)

The 101-key keyboard has four key groups: first, the **function key** row at the top of the keyboard; second, the main (typewriter) **keypad**; third, the **cursor-edit keypad** with arrows indicating up, down, right, and left directions for moving the cursor; and fourthly, the **numeric** (calculator style) **keypad**, which has two operating modes (when the Num Lock indicator light is on, the keypad is in the numeric mode; to switch the keypad to the editing mode, press the **Num Lock key** to turn off the Num Lock indicator) .

在大多数计算机上，键盘是主要的文本输入设备。（鼠标也是主要的输入设备，但在录入文本信息方面却能力有限。）

101 键位的键盘有 4 个键区：第一，键盘上部的功能键区；第二，主键区（打字键区）；第三，光标/编辑键区，包括上、下、左、右四个箭头键，用于移动光标；第四，数字键区（计算器式的），具有两种工作模式（当 Num Lock 指示灯亮时，该键区处于数字输入状态；要转换到编辑状态，只需按 Num Lock 键，关闭 Num Lock 指示灯即可）。

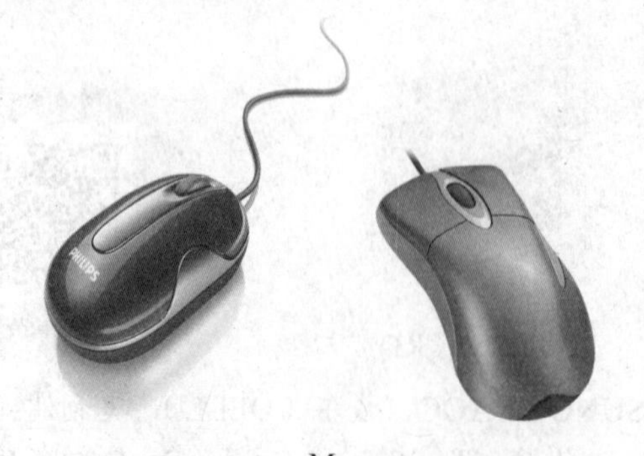

Mouse

 **Key words:** function key（功能键），keypad（小键盘），cursor-edit keypad（光标-编辑键盘），numeric keypad（数字键盘），Num Lock key（数字锁定键）

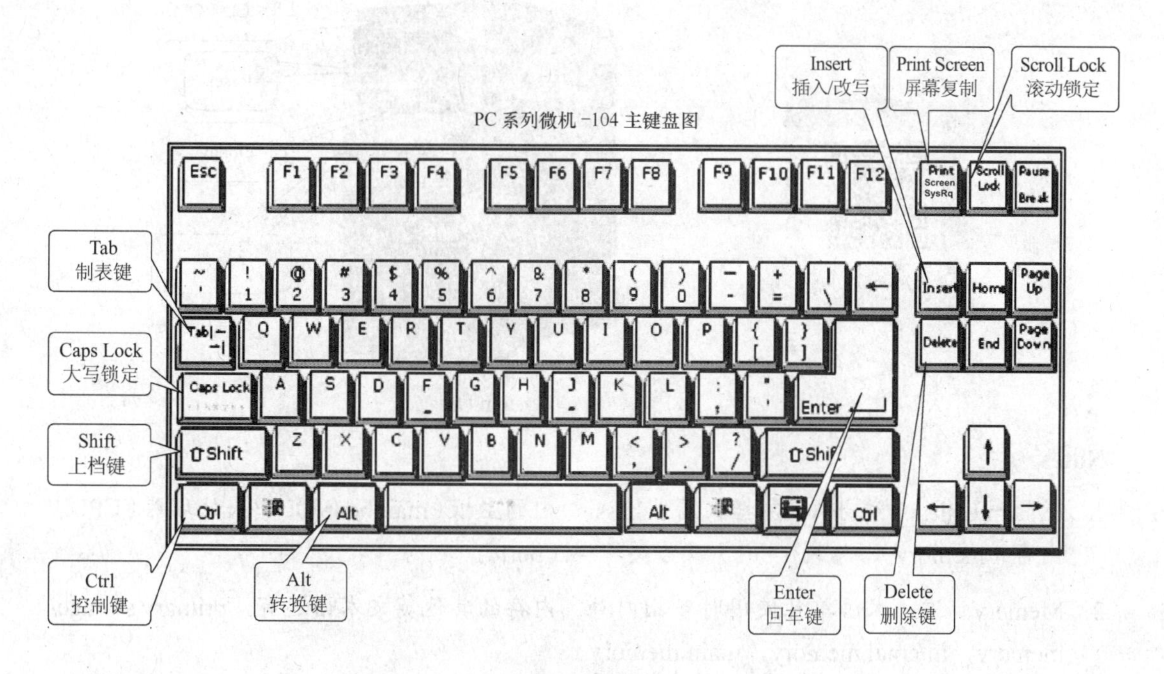

1.2.3 System Unit（主机）

The system unit is the electronic circuitry housed within the computer cabinet. The two main parts of the system unit are:

1. The central processing unit (CPU), which controls and manipulates data to produce **information**.

2. **Memory**, also known as **primary storage**, which holds data and program **instructions** for processing the data.

系统部件是安装在计算机机箱中的电子线路的总称。系统部件的主要的两部分是：

1．中央处理器（CPU），控制和处理数据以输出信息。

2．内部存储器，也就是主存，保存数据及处理数据的程序指令。

System Unit

Notes

1. System Unit: 系统部件，包括机箱（chassis）内的主板（mainboard）、中央处理器（CPU）、内存（memory）、总线（bus）及各类端口（port）。
2. Memory: 这一术语单独使用时多指内存。内存的其他英文术语还有：primary storage/memory，internal memory，main memory。

Key words: information（信息），memory（内存），primary storage（主存），instruction（指令）

1.2.4 Printer（打印机）

A **printer** is a device that accepts text and graphic output from a computer and transfers the information onto paper，usually to **standard-sized sheets** of paper.

打印机是一种从主机接收文本或图形信息，并将其传送到打印纸上的设备，通常使用标准纸张。

Personal computer printers can be classified as **impact** and **non-impact printers**. The **dot-matrix printer** has been a popular low-cost personal computer printer. It's an **impact printer** that strikes the paper a line at a time. The best-known non-impact printers are the **inkjet printer**, of which several makes of low-cost **color printers** are examples, and the **laser printer**.

个人计算机配套的打印机可分成击打式和非击打式两种。点阵式打印机价格较低，一直是个人计算机上流行的打印机。点阵式打印机属击打式的，每次在纸上打印一行。最常见的非击打式打印机当属喷墨打印机（比如市场上那几款低价位的彩色喷墨打印机）和激光打印机。

常见的打印机品牌：佳能（Canon）、惠普（HP）、爱普生（Epson）、富士施乐（Fujixerox）、三星（SAMSUNG）、兄弟（Brother）、联想（Lenovo）、OKI 等。

Key words: printer（打印机），standard-sized sheets（标准页），impact printer（击打式打印机），non-impact printer（非击打式打印机），dot-matrix printer（点阵打印机），inkjet printer（喷墨打印机），color printer（彩色打印机），laser printer（激光打印机）

Laser Printer

Inkjet Printer

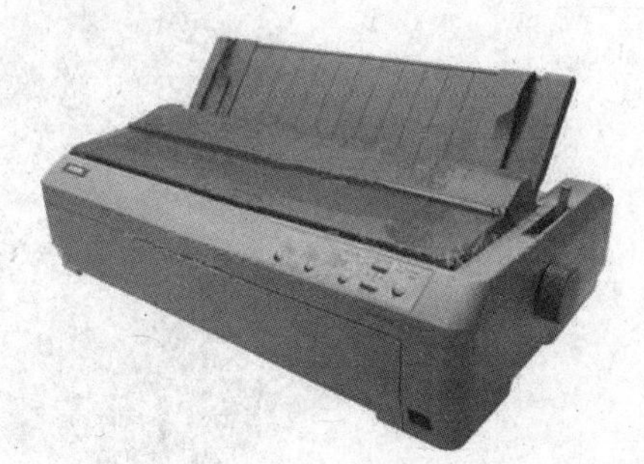

Dot-matrix Printer

1.3 Computer Components（计算机组件）

计算机组件主要包括：

Mainboard/System board（主板），Memory（内存），CPU（中央处理器），Video adapter（视频适配器），Hard disk（硬盘），Multimedia（多媒体）。

1.3.1 System Board（主板）

System boards are almost as important as CPUs in computer system. If a CPU were a heart or a brain, a system board would be a blood vessel or nerve system. A CPU controlls and manages the whole system with the help of a system board.

A system board actually is the largest **circuit board** among computer components. There are many electronic elements, **sockets**, **slots**, and connects on it, which link CPU with **peripherals** together.

在计算机系统中，主板几乎和CPU一样重要。如果CPU是心脏或大脑，那么主板则是血管或神经系统。CPU必须在主板的帮助下来控制和管理整个系统。

主板实际上是计算机部件中最大的电路板。在它上面，有许多元件、插座、插槽和接插件，它能把CPU和外围设备连接在一起。

Key words: system board（主板），circuit board（电路板），socket（插座），slot（插槽），peripheral（外围设备）

常见的主板品牌：华硕（ASUS）、技嘉（GIGABYTE）、微星（msi）、映泰（BIOSTAR）、华擎（Asrock）、精英（ECS）、七彩虹（Colorful）、昂达（ONDA）、梅捷（SOYO）、英特尔（Intel）等。

PCI：Peripheral Component Interconnect 外围组件互联（一种局部总线标准）
PCIE：PCI Express 高速的 PCI
USB：Universal Serial Bus 英特尔公司开发的通用串行总线架构
IDE：Integrate Circuit Equipment 集成电路设备
AGP：Accelerated Graphics Port 高速图形接口

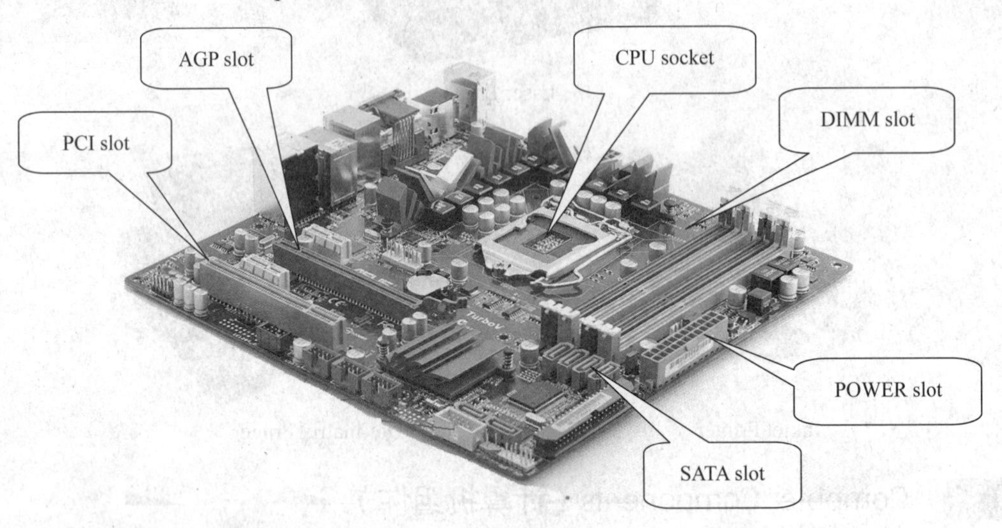

1.3.2 Memory（内存）

Your personal computer comes with a minimum of 8 **gigabytes** of memory or even more, depending on your system.

个人计算机一般都带有8GB甚至更大的内存，具体的内存量视系统而定。

System memory is located on chips inside the computer, and is measured in **bytes** (the amount of storage needed to hold one **character**).

系统内存位于机器内部的集成电路芯片中，其容量以字节度量（一个字节也就是存储一个字符所占用的存储空间）。

There are categories of system memory: **Read Only Memory** (ROM), and **Random Access Memory** (RAM). ROM contains programs and data that never change. This is the memory that **initializes** the computer when it is **turned on**. The information contained in ROM is permanent, and is not lost when the computer is **turned off**. By contrast, RAM is temporary storage for programs and data while they are being used by the computer. RAM is contained in memory components called Single Inline Memory Modules or **SIMMs**. It is volatile memory, which means that to retain its contents, it must be constantly **refreshed** by an electrical current.

系统内存可分为两类：只读存储器（ROM）及随机存储器（RAM）。ROM 用以存储固定的程序和数据。计算机开机后，ROM 中的程序负责对系统进行初始化。ROM 中的信息是永久性的，即使计算机关机后也不会丢失。相比之下，RAM 则是用于暂时存储计算机正在使用的程序和数据的。RAM 位于叫作内存条（SIMM）的存储部件中。RAM 是易失性的，也就是说，为了保持其中的内容，必须由电信号经常进行刷新。

Key words: gigabyte（吉字节），byte（字节），character（字符），ROM（只读存储器），RAM（随机存储器），initialize（初始化），turn on（开机），turn off（关机），SIMM（内存条），refresh（刷新）

ValueRAM DDR3 1333 台式机内存

DIMM(Dual In-line Memory Module)

常见的内存厂商：金士顿（Kingston）、威刚（ADATA）、海盗船（CORSAIR）、三星（Samsung）、宇瞻（Apacer）、芝奇（G.SKILL）、十铨（TEAM）、海力士（HYNIX）、英睿达（CRUCIAL）、金邦科技（GEIL）等。

Notes

SIMM：Single In-line Memory Module，指 72 线的内存条，数据通道宽度为 32 位，带有奇偶校验的则为 36 位，通常使用 DRAM；也有的 SIMM 使用 SDRAM，其速度比 DRAM 类型的更快，但由于其数据宽度为 64 位，需成对使用。另外，目前多使用 168 线的内存条，称为 DIMM（Dual In-line Memory Module），其数据宽度是 64 位，一条 DIMM 内存条相当于两条 SIMM，所以可单条使用。

1.3.3 CPU（Central Processing Unit，中央处理器）

The CPU means the central processing unit. It is the heart of the computer system. The CPU in a microcomputer is actually one relatively small **integrated circuit** or chip. Although most CPU chips are smaller than a lens of a pair of glasses, the electronic components they contain would have filled a room a few decades ago. Using advanced microelectronics techniques, manufacturers can cram tens of thousands of circuits into tiny layered silicon chips that work dependably and use less power.

CPU 就是中央处理器的意思。它是计算机的心脏。微型计算机上的 CPU 实际上是一个不大的集成电路芯片。虽然大多数 CPU 芯片比一块眼镜片还小，但在几十年前，它们所容纳的电子元器件能装满一个房间。然而，采用先进的微电子技术，制造者们能够把成千上万个电子元件集成到很小很薄的硅片上，这些硅片的工作性能可靠且不费电。

This is where all calculations and manipulations of the data are carried out: it could be considered the "brain" of the computer. The CPU is contained on a tiny integrated circuit or "microchip" that carries a large number of minute electronic circuits.

中央处理器是执行数据操作和运算的地方，它可以被看成计算机的大脑。中央处理器装在一个极小的集成电路或微集成电路里。这个微集成电路装有大量的极小的电子电路。

There are three main areas in the CPU: the **control unit**, the **arithmetic logic unit (ALU)**, and the main storage of "memory".

中央处理器有 3 个主要区域：控制部件、算术逻辑部件及主存储器。

Key words: integrated circuit（集成电路），control unit（控制单元），ALU（运算器）

AMD is another CPU brand, which has become very important. Their Pentium-like chips offered Intel tight competition. AMD used their own technologies, hence they are not clones.

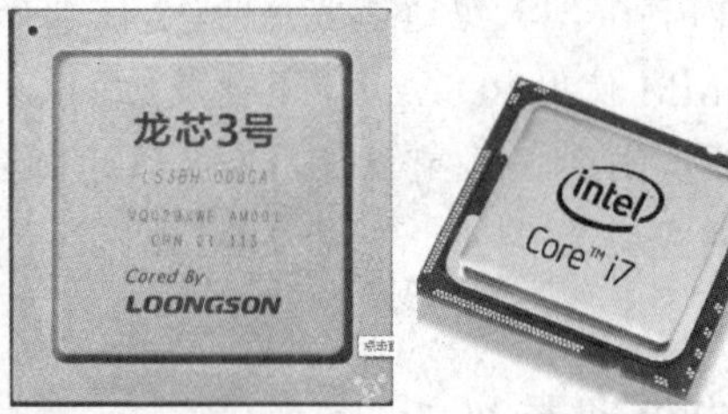

常见的 CPU 生产厂商：英特尔公司（Intel）、超微半导体公司（AMD）、龙芯（LOONGSON）。

Notes

CPU：Central Processing Unit，中央处理器，在大型计算机上 CPU 是由主存储器、控制部件、算术逻辑部件等若干单元电路板构成的。在个人计算机上，CPU 是集成于一块芯片上的，称作微处理器（microprocessor 或 processor）。通常，CPU 和 microprocessor 或 processor 可以互换使用。CPU 主要有两大部分：算术—逻辑部件（Arithmetic Logic Unit，ALU）负责算术及逻辑运算；控制部件（Control Unit）负责指令的存取、解码及执行。

1.3.4 Video Adapter（视频适配器）

Video adapters are also called **video cards**, video boards, video display boards, **graphics cards**, and graphics adapters.

视频适配器又叫显卡、视频板、视频显示板、图形卡或图形适配器。

Video adapter is a board that plugs into a personal computer to give it display capabilities. The display capabilitiy of a computer, however, depends on both the **logical circuitry** (provided in the video adapter) and the display monitor. A **monochrome monitor**, for example, cannot display colors no matter how powerful the video adapter is.

Many different types of video adapters are available for PCs. Most conform to one of the video standards defined by IBM or VESA.

Modern video adapters contain memory, so that the computer's RAM is not used for storing displays. In addition, most adapters have their own graphics **coprocessor** for performing graphics calculations. These adapters are often called **graphics accelerators**.

视频适配器是为个人计算机提供显示功能的一块插板。当然，计算机要有显示功能必须同时具备（视频适配器提供的）逻辑电路和显示器。例如，无论视频适配器功能多强大，一个黑白显示器终究还是不能显示彩色的图形。

许多不同类型的视频适配器都是与 PC 兼容的。绝大多数都遵循由 IBM 或 VESA（视频电子标准协会）制定的视频标准。

现在的视频适配器都有自己的内存，因此计算机的内存就不用存储显示内容了。另外，许多适配器还有了自己的图形协处理器，用以执行图形计算，这些适配器通常被称为图形加速器。

Key words: video adapter（视频适配器）, video card（显卡）, graphics card（显卡）, logical circuitry（逻辑电路）, monochrome monitor（单色显示器）, coprocessor（协处理器）, graphics accelerator（图形加速器）

常见显卡品牌：七彩虹（Colorful）、蓝宝石（Sapphire）、影驰（GALAXY）、华硕（ASUS）、微星（msi）、技嘉（GIGABYTE）、铭瑄（MAXSUN）、迪兰（Dataland）、索泰（ZOTAC）、丽台（Leadtek）等。

Video Card

1.3.5 Hard Disk（硬盘）

The **hard disk** drive in your system is the "**data center**" of the PC. It is here that all of your programs and data are stored between the occasions that you use the computer. Your hard disk (or disks) are the most important ones of the various types of **permanent storage** used in PCs (the others being **floppy disks** and other storage media such as **CD-ROMs**, **tapes**, removable drives, etc.) The hard disk differs from the

个人计算机的硬盘是整个系统的"数据中心"，它是存储计算机中所有程序和数据的地方。硬盘在 PC 所使用的各种永久性存储器中（如光盘、磁带、移动驱动器等）是最重要的一种。它的主要不同之处在于 3 个方面：容量（通常更大），速度（通常更快）和固定性（通常被

others primarily in three ways: size (usually larger), speed (usually faster) and permanence (usually fixed in the PC and not removable).

固定在 PC 中而不能移动）。

A single hard disk usually consists of several platters. Each platter requires two read/write heads, one for each side. All the read/write heads are attached to a single **access** arm so that they cannot move independently. Each platter has the same number of **tracks**, and a track location that cuts across all platters is called a **cylinder**.

一个硬盘通常有多个盘片。每个盘片又有两个读/写磁头，一面一个。所有的读/写磁头都被固定在一个磁头臂上，因此它们不能各自独立地移动。每个盘片都有相同数量的磁道，所有盘片的（与其垂直相切的）同一磁道位置构成一个柱面。

With the development of technology, Solid State Drives(SSD)arises at the historic moment. A solid-state drive is a data storage device using integrated circuit assemblies as memory to store data persistently. SSDs have no moving mechanical components. This distinguishes them from traditional electromechanical magnetic disks such as hard disk drives (HDDs) or floppy disks. SSDs have the same function, definition and method of use as Hard Disk Drive (HDD). Compared with HDDs, SSDs are typically more resistant to physical shock, run silently, and have lower access time and lower latency, lower noise performance, handiness, wide temperature sensitive and so on. But SSDs are expensive. They have relative smaller capacity and shorter lifespan. At present, the mechanical hard disk storage capacity can reach T level. But now the mainstream capacity of SSD is 120G or 240G, of course, there are also 500G. Of course, SSDs have the capacity of 1T, but the high price is not acceptable to ordinary users.

随着技术的发展，固态硬盘（简称固盘）应运而生。固态硬盘是用固态电子存储芯片阵列而制成的存储设备。固态硬盘没有移动的机械磁头，这是它们和传统的机械磁盘如 HDD 和软盘的区别。固态硬盘与机械硬盘有相同的功能、定义和使用方法。与普通硬盘相比，固态硬盘抗震性能好，功耗低，反应快，读写速度快，无噪声，轻便，工作温度范围大等特点，但价格昂贵，容量相对较小，使用寿命要短些。目前机械硬盘存储容量能达到 T 级别，而现在固态硬盘的主流容量是 120G 或者 240G，当然也有 500G。当然，固态硬盘也有 1T 容量的，但是那高高在上的价格是一般用户接受不了的。

Hard Disk

SSD

常见硬盘品牌：希捷（Seagate）、西部数据（Western Digital）、日立（Hitachi）、三星（SAMSUNG）、东芝（TOSHIBA）、联想（ThinkPad）、IBM、富士通（FUJITSU）、英特尔（Intel）、英睿达（Crucial）、金士顿（Kingston）等。

Key words: hard disk（磁盘），data center（数据中心），permanent storage（永久性存储器），floppy disk（软磁盘），CD-ROM（只读光盘），tape（磁带），access（存取），track（磁轨），cylinder（柱面）

1.3.6 Multimedia（多媒体）

Probably, several years ago, you may say proudly: "I've got a **multimedia** computer". Actually you meant that you had a wonderful multi-function PC. You can not only process files or data, but furthermore can listen to music, watch cartoons, play games or even talk to the computer. That's the idea of the buzzword multimedia. Now it has been the basic **technology** adopted on every computer.

In fact, multimedia is a kind of technology, or more accurately a combination of technologies. It's not a simple **product**, like your computer. It involves text, audio sound, static graphic images, animations, and full-motion video. Multimedia may use some or all of these aspects of communication. Text in a multimedia **application** makes it more understandable, and can display information related to a certain topic. **Animation** refers to moving **graphics images**, which is more vivid to illustrate concepts concerning movements than static graphics images, which provide many "still **pictures**". **Audio sound** can be of several formats: Windows wave file, Musical Instrument Digital Interface, or **MIDI** for short. Full-motion **video** refers to the video got from an

也许几年前，你可以骄傲地说："我有一台多媒体计算机。"因为你不但可以进行文本、数据处理，还可以听音乐、看动画、玩游戏，甚至可以进行人机对话。的确，计算机为你提供了多功能享受，这就是流行词"多媒体"背后的概念。现在多媒体技术已经普遍成为计算机的基本配置。

多媒体实际上是一种技术，更准确地说是多种技术的综合，而非产品，比如你的家用计算机。它是文本、声音、静态图像、动画和全运动图像的集成。多媒体可以部分或全部地使用这些交流方式。多媒体中的文本使其更容易理解，并且能显示与主题相关的信息。动画指移动图像，它比只提供静止图片的静态图像更能生动地说明与运动相关的概念。声音可以是几种不同的格式，如 Windows 声波文件、乐器数字接口（MIDI）。全运动图像是指通过外部输入，如摄像机，并存储在硬盘上的图像。它给多媒体应

external input, like a **video camera**, and stored on the hard disk. It adds powerful message to multimedia application. These are the basic elements of multimedia. Every combination of two or more elements makes a multimedia application.

用增加了非常丰富的信息。这些就是多媒体的基本组成要素，凡具备两种或两种以上要素的就是多媒体。

 **Key words:** multimedia（多媒体），technology（技术），product（产品，产物），application（应用，应用程序，应用软件），animation（动画），graphics image（图形图像），picture（图画），audio（音频 n，声音的 adj），sound（声音），MIDI（数码音响），video（视频 n，影像的 adj），video camera（视频照相机）

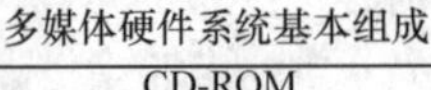

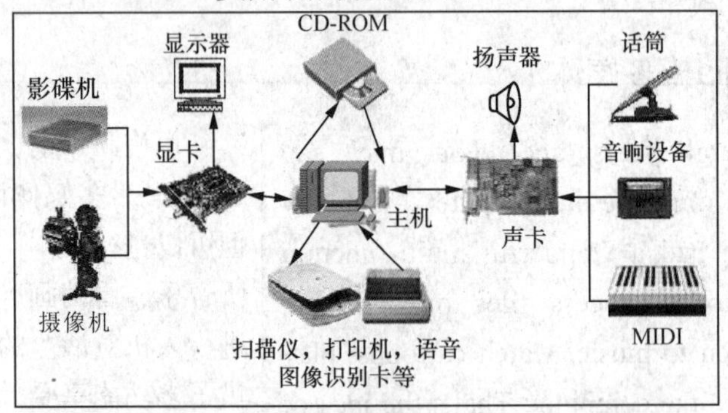

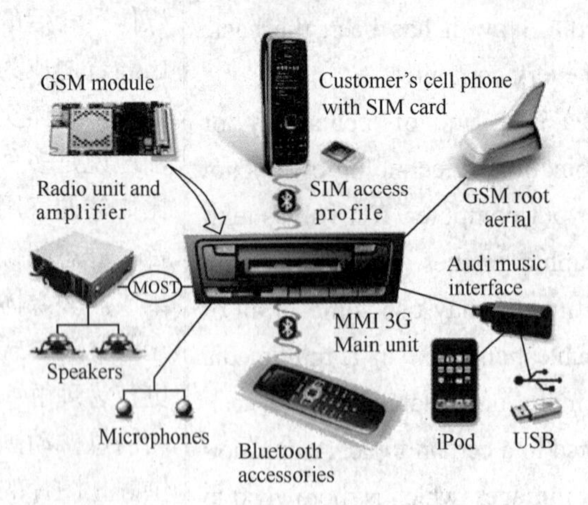

 多媒体外部设备如下。

视频、音频输入设备：摄像机（vidicon）、录像机（video cassette recorder）、扫描仪（scanner）、传真机（electrograph）、数字相机（figure camera）、话筒（mike）等。

人机交互设备：键盘（keyboard）、鼠标（mouse）、触摸屏（touch screen）、绘图板（drawing table）、光笔（lightpen）及手写输入设备（input equipment）等。

1.4 Situation Dialogue（情境对话）

Visiting the Computer Supermarket

Miss Li: a salesgirl in Computer Town

Mr. Zhang: a computer consumer who knows little about computer

Li: Can I help you, Sir?

Zhang: Yes. I'd like to buy a computer, but I know little about computer.

Li: Do you like a brand machine or a compatible machine?

Zhang: But what is a brand machine?

Li: A brand machine is one that is designed and assembled by some famous computer companies in the world. It is tested with great care after being assembled. It has good quality and has good service after sale.

Zhang: What kind of brands do you have, please?

Li: IBM, HP, DELL abroad. Domestically, we have Lenovo, Founder, Tsinghua Tongfang, TCL, Hasee, and the like.

Zhang: What is a compatible computer, then?

Li: A compatible is also called DIY (Do It Yourself). It is assembled by the user. It can be very cheap.

Zhang: OK. I'd like to buy a compatible computer. Could you give me some suggestions?

Li: Sure, sir! In my opinion, I suggest you dispose your computer as followings:

Mainboard: ASUS P7H55-M Pro.

CPU (central processing unit): Intel Core i5 750.

Memory: Kingston 2GB DDR3 1333.

Hard Disk:WD500G/7200/16M/serial port (WD 5000AA).

逛计算机超市

李小姐：电脑城销售员

张先生：对计算机不太懂的顾客

李小姐：先生，请问有什么可以帮您的吗？

张先生：是的，我想买计算机，但我不怎么了解计算机。

李小姐：先生，您喜欢品牌机还是兼容机？

张先生：什么叫品牌机？

李小姐：由世界上一些著名计算机公司设计、组装的计算机。出厂前已经过严格测试，质量好，售后服务好。

张先生：请问有哪些品牌？

李小姐：国外的有 IBM、惠普、戴尔，国内的有联想、方正、清华同方、TCL，还有神舟等。

张先生：那么小姐，请问什么叫兼容机？

李小姐：兼容机又称 DIY。它是由用户自己组装的，价格便宜。

张先生：那我想买一台兼容机，你能给我点建议吗？

李小姐：先生，根据经验，我建议您做这样的配置：

主板：华硕 P7H55-M Pro

CPU：英特尔 酷睿 i5 750

内存：金士顿 2GB DDR3 1333

硬盘：西部数据 500GB/7200 转/16MB/串口（WD5000AA）

Video Card: Colorful iGame5770-GD5.

Disk Driver: Pioneer DVR-218CHV.

Monitor: AOC iF23.

Printer: Lenovo LJ1680Home Edition.

As for Computer chassis and speaker, it's up to you.

Zhang: Can you guarantee the service after sale?

Li: Certainly.

Zhang: OK. Thank you very much.

Li: You are welcome.

显卡：七彩虹 iGame5770-GD5

光驱：先锋 DVR-218CHV

显示器：AOC iF23。

打印机：联想 LJ1680 家庭版

机箱和音箱您可以自己任意挑选。

张先生：售后服务有保障吗？

李小姐：当然有，请您放心使用。

张先生：好的，谢谢。

李小姐：不客气。

朗读音频

Situation Dialogue

1.5 Reading and Compacting（对照阅读）

Computer Systems

A computer is a fast and accurate symbol manipulating system that is organized to accept, store, and process data and produce output results under the direction of a stored program of instructions. A complete computer set is a system that basically includes the following four key parts: the processor, the memory, the Input and the Output.

The processor is the "brain" of the computer that has the ability to carry out our instructions or programs given to the computer. The processor is the part that knows how to add and subtract and to carry out simple logical operations. In a big mainframe computer the processor is called a Central Processing Unit, or CPU, while in a microcomputer, it is usually known as a microprocessor.

The memory is the computer's work area and nothing like our own memory, so the term can be misleading. The computer's memory is where all activities take place. The size of a computer's memory sets a practical limit on the kinds of work that the computer can undertake.

计算机系统

计算机是一个快速和精确的符号处理系统，它被组织成能在所存储程序指令的作用下接受、存储和处理数据以及产生输出结果。一个完整的计算机系统基本上包括以下 4 个部分：处理器、存储器、输入部分和输出部分。

处理器是计算机的“大脑”，它具有执行用户提交给系统的指令或程序的能力。处理器是实现加法和减法运算以及执行简单逻辑运算的地方。在大型计算机系统中，处理器被称作中央处理器，或 CPU，而在微型计算机系统中，通常叫作微处理器。

存储器是计算机的工作区，但它完全不像我们的记忆系统，所以这个术语可能会引起误解。计算机的存储器是所有活动发生的场所。计算机存储器的容量限制了计算机能承担的工作的种类。

I/O are all the means that the computer uses to take in or out data. It includes input that we type in on the keyboard and output that the computer shows on the video display screen or prints on the printer. Every time the computer is taking in or putting out data, it's doing I/O using I/O devices, which is also called peripheral devices.

I/O 是计算机用来获取数据或输出信息的所有手段。它包括我们借助于键盘的输入和通过显示器或打印机的输出。每次计算机输入或输出数据，都得通过被称作外围设备的 I/O 设备实现 I/O 操作。

Most of the input devices work in similar ways. The messages or signals received are encoded into patterns which CPU can process by input devices, then, conveyed to CPU. Input devices can not only deliver information to CPU but also activate or deactivate processing just as light switches turn lamps on or off.

大部分输入设备以相似的工作方式工作。输入设备把接收到的信息和信号进行加工，使之变成 CPU 能处理的编码，然后传送给 CPU。输入设备不仅能给 CPU 输送信息，而且也能像开关控制电灯那样激活或终止处理过程。

Output devices also play important roles in a computer system. They can tell the processing results and warn users where their programs or operations are wrong. The most common output devices are monitor, matrix printer, inkjet printer, laser printer, plotter for drawing, speaker, etc. They also work in similar ways. They decode the coded symbols produced by CPU into forms of information that users understand or use easily and show them.

输出设备在计算机系统中也扮演着重要的角色。它们能告诉用户处理的结果和警告用户程序或操作在哪里错了。最普遍的输出设备有显示器、点阵打印机、喷墨打印机、激光打印机、绘图仪和喇叭等。它们也以相似的方式工作：把 CPU 产生的编码形式的信号解码，变成人们易懂和易用的形式并显示出来。

Input · Output · Storage · Processing

How They Work Together

First, you provide input when you turn on the computer. Then the system software tells the CPU to start up certain programs and to turn on some hardware devices, so that they are ready for more input from you. This whole process is called booting up.

The next step happens when you choose a program you want to use. You click on the icon or enter a command to start the program. Let's use the example of an Internet browser. Once the program has started, it is ready for your instructions. You either enter an address (called a URL, which stands for Uniform Resource Locator), or click on an address you've saved already. In either case, the computer now knows what you want it to do. The browser software then goes out to find that address, starting up other hardware devices, such as a modem, when it needs them. If it is able to find the correct address, the browser will then tell your computer to send the information from the web page over the phone wire or cable to your computer. Eventually, you see the web site you were looking for.

If you decide you want to print the page, you click on the printer icon. Again, you have provided input to tell the computer what to do. The browser software determines whether you have a printer attached to your computer, and whether it is turned on. It may remind you to turn on the printer, then send the information about the web page from your computer over the cable to the printer, where it is printed out.

它们如何协调工作

首先，开机时会提供一个输入信号。然后，系统软件会告诉 CPU 去运行某个程序并启动一些硬件，以使系统准备接收用户输入的信息。这整个过程称之为系统引导。

接下来的一步是选择你想要运行的程序，单击应用程序的图标或是输入一个命令都可启动程序。以浏览器为例，一旦程序启动，就可以接受指令了。你既可以输入一个地址（即 URL，统一资源定位），也可单击收藏夹里的网址。不论哪一种，计算机都知道该做什么了。浏览器会自动地搜索地址。如果需要的话，可以启动其他的硬件，如调制解调器。若能找到正确的网络地址，计算机会通过电话线或电缆将网页上的信息在浏览器中显示出来。最后，就能看到所搜索的网站了。

如果你决定要打印当前页，只需单击打印图标。这样，也就再一次提供了一个告诉计算机做什么的输入。浏览器会检测计算机是否连有打印机，并检测打印机是否是打开的。计算机也许会提醒你打开打印机，然后通过电缆把网页上的信息传递给打印机，于是这些内容就打印出来了。

1.6 术语简介

1. VGA：Video Graphics Array，视频图形阵列，是 1987 年由 IBM 公司推出的一种图形显示技术，事实上已经成为业内标准。

2. Serial Port：串行口，主要用于连接鼠标、调制解调器、终端等低速通信设备。

3. Parallel Port：并行口，主要用于连接打印机、扫描仪等高速通信设备。

4. KB：Kilo- + Byte，表示 1024 个字节。类似的希腊语前缀还有 Mega-(10^6)、Giga-(10^9)、

Tera-(10^{12})，与 Byte 组词后一般缩写为 KB、MB、GB、TB。

5．ROM：Read Only Memory，只读存储器，用以存储 BIOS 等信息，用户不能随意修改。

6．RAM：Random Access Memory，随机存储器，人们通常所说的内存就是指 RAM。

Exercises（练习）

1．Match the explanations in Column B with words and expressions in Column A.（搭配每组中意义相同的词或短语）

A	B	A	B
打印机	Modem	显示	Computer
驱动器	Printer	故障	Write protection
元件	Hardware	计算机	Mouse
磁盘	Pathname	键盘	Display
路径名	Component	写保护	LAN
硬件	Disk	鼠标	Fault
调制解调器	Drive	局域网	Keyboard

2．Choose the proper words to fill in the blanks.（选词填空）

Types of Computer

Computer can be generally classified by size and power as following, though there is considerable overlap among them.

A ____________ refers to a small, single-user computer based on a microprocessor, with a keyboard for entering data, a monitor for displaying information, and a storage device for saving data.

A ____________ is a powerful, single-user computer, which, though looking like a microcomputer, has a more powerful microprocessor and a higher-quality monitor.

A ____________ stands as a multi-user computer capable of supporting from 10 to hundreds of terminals（终端）or users simultaneously（同时地）.

A ____________ , synonymous with "giant", is the name for a powerful multi-user computer which supports hundreds of thousands of users at the same time.

A ____________ describes an extremely fast computer that can perform hundreds of millions of instructions per second.

Words to be chosen from:（可选词）

Supercomputer, minicomputer, mainframe, workstation, personal computer

3．Translate the following English phrases or sentences into Chinese.（将下面英语短语或句子翻译为中文）

Personal Computer.

This is a display. The display is on the system unit.

That is a printer. The printer is by the display and system unit.

This is a disk drive. The disk drive is in the system unit.

This is a keyboard, and that is a mouse. The keyboard and mouse are before the system unit and display.

That is a modem. The modem is by the system unit, too.

They are basic hardware components of a personal computer system.

4．实践作业：到当地电脑城了解计算机硬件的最新动态和报价。

附文1: Reading Material（阅读材料）

PC Overview

A PC is a small, relatively inexpensive computer designed for an individual user. In price, personal computers range anywhere from a few hundred dollars to over five thousand dollars. All are based on the microprocessor technology that enables manufacturers to put an entire CPU on one chip. Businesses use personal computer for word processing, accounting, desktop publishing, and for running spreadsheet and database management applications. At home, the most popular use for personal computers is for playing games and computer-assisted learning.

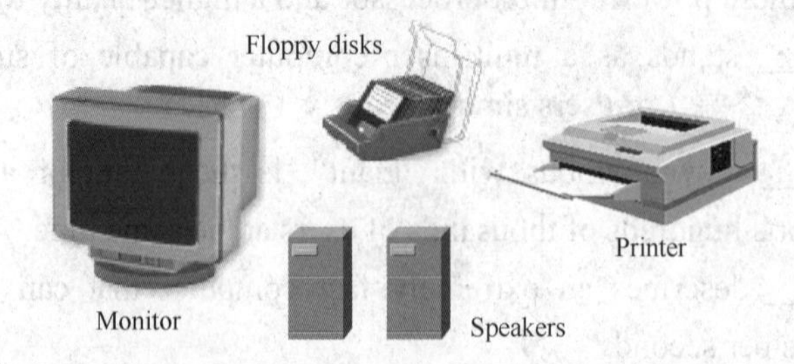

Personal computers first appeared in the late 1970s. One of the first and most popular personal computers was the Apple Ⅱ, introduced in 1977 by Apple Computer. During the late 1970s and early 1980s, new models and competing operating systems seemed to appear daily. Then, in 1981, IBM entered the fray (竞争) with its first personal computer, known as the IBM PC, the IBM PC quickly became the personal computer of choice, and most other personal computer manufacturers fell by the wayside. One of the few companies to survive IBM's onslaught (冲击) was Apple Computer, which remains a major player in the personal computer marketplace.

Other companies adjusted to IBM's dominance (优势) by building IBM clones, computers that were internally almost the same as the IBM PC, but that cost less. Because IBM clones used the same microprocessors as IBM PCs, they were capable of running the same software. Over the years, IBM has lost much of its influence (影响) in directing the evolution (发展) of PCs. Many of its innovations (创新), such as the MCA (Micro Channel Architecture) expansion bus and the OS/2 operating system, have not been accepted by the industry or the marketplace.

Today, the world of personal computer is basically divided between Apple Macintoshes and PCs. The principal (主要的) characteristics (特征) of personal computers are that they are single-use systems and are based on microprocessors. However, although personal computers are designed as single-user systems, it is common to link them together to form a network. In terms of power, there is a great variety. At the high end, the distinction between personal computers and workstations has faded (已褪色的). High-end models of the Macintosh and PC offer the same computing power and graphics capability as low-end workstation by Sun Microsystems, Hewlett-Packard, and DEC.

How Do Computers Work?

To accomplish a task using a computer, you need a combination of hardware, software and input.

Hardware consists of devices, like the computer itself, the monitor, keyboard, printer, mouse, and speakers. Inside your computer there are more bits of hardware, including the motherboard, where you would find the main processing chips that make up the central processing unit (CPU). The hardware processes the commands it receives from the software, and performs tasks or calculations.

Software is the name given to the programs that you install on the computer to perform certain types of activities. There is operating system software, such as the Apple OS for a Macintosh, or Windows 95 or Windows 98 for a PC. There is also application software, like the games we play or the tools we use to compose letters or do math problems.

You provide the input. When you type a command or click on an icon, you are telling the computer what to do. That is called input.

Chapter 2 Computer System Maintenance（计算机系统维护）

教学要求

本章是计算机维护的必备知识，要求学生对硬件已有了解。

教学内容

开机全过程，CMOS，设备添加、删除与配置，常用 DOS 命令与网络命令。

教学提示

建议到机房上课，特别是 DOS 命令和网络命令，通过练习才能掌握。

2.1 Booting the Computer(启动计算机)

Whenever you turn on your computer, the first thing you see is the **BIOS** (Basic Input/Output System) software which is running. On many machines, the BIOS displays text describing things like the amount of memory **installed** in your computer, the type of hard disk and so on. It turns out that, during this **booting**, the BIOS is doing a **remarkable** amount of work to get your computer ready to run. This section **briefly** describes some of those activities for a typical PC.

启动计算机，你所看到的第一件事就是 BIOS（基本输入输出系统）软件正在执行它的操作。在许多计算机上，BIOS 会显示出文字描述信息，比如计算机内存容量、硬盘类型等。这意味着在启动的程序中 BIOS 正在进行大量的工作以使你的计算机做好运行前的准备。这一节我们以常见的计算机为例简要地介绍计算机的启动过程。

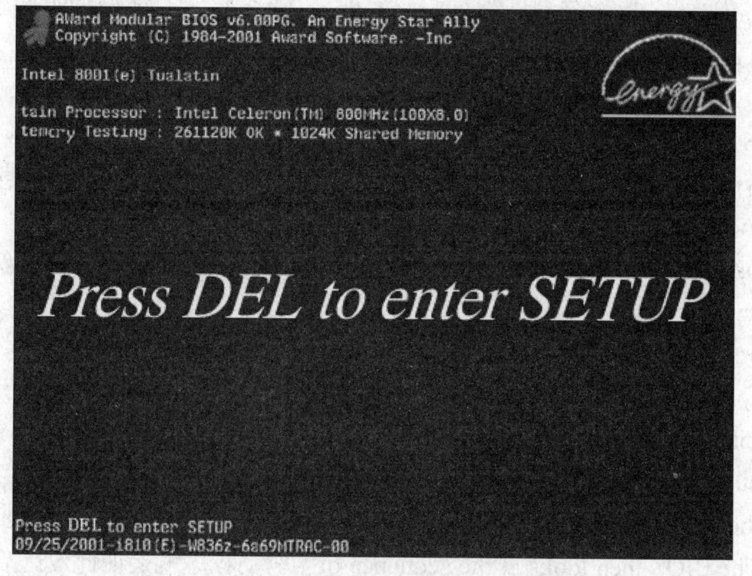

Boot the Computer

Key words: BIOS（基本输入输出系统）, install（安装，安置）, boot（启动）, remarkable（不同寻常的）, briefly（暂时地，简要地）

After checking the **CMOS** Setup and loading the **interrupt handlers**, the BIOS determines whether the video card is operational. Most video cards have a **miniature** BIOS of their own that initializes the memory and **graphics** processor on the card. If they do not, there is usually video driver information on another **ROM** on the motherboard that the BIOS can load.

检查 CMOS 设置、载入中断处理程序后，BIOS 检测显卡是否工作。大多数的显卡都自带有微型 BIOS 来初始化显卡上的显存和图形处理器。如果显卡没有自带 BIOS，通常在主板的另一个只读存储器中会存有 BIOS 能够载入的显卡驱动程序。

Next, the BIOS checks to see if this is a cold boot or a reboot. It does this by checking the value at memory address 0000:0472. A value of 1234h indicates a reboot, and the BIOS skips the rest of **POST**. Anything else is considered a cold boot.

紧接着，BIOS 通过检测内存地址 0000:0472 中的值来检查此次启动是冷启动还是热启动。若该内存地址中的值为 1234h 表示热启动，则 BIOS 将跳过剩下的加电自检程序。检测到任何其他的值 BIOS 都将认为此次启动是冷启动。

If it is a cold boot, the BIOS verifies **RAM** by performing a read/write test of each memory address. It checks the PS/2 ports or **USB** ports for a keyboard and a mouse. It looks for a **peripheral component interconnect** (**PCI**) bus. If it finds one, it checks all the PCI cards. If the BIOS finds any errors during the POST, it will notify you by a series of beeps or a text message displayed on the screen. An error at this point is almost always a hardware problem.

如果是冷启动，BIOS 通过对每个内存地址进行读写测试来校验随机存储器。BIOS 检查键盘和鼠标的 PS/2 端口或者 USB 端口。然后 BIOS 查找互连外围设备（PCI）总线。如果它找到一个，则检测所有的 PCI 卡。如果 BIOS 在执行自检程序过程中发现任何错误，则会通过一系列的鸣叫声或在屏幕上显示文字信息来提示你。此时的错误几乎都是硬件方面的问题。

The BIOS then displays some details about your system. This typically includes information about:

- The processor
- The floppy driver and hard driver
- Memory
- BIOS revision and date
- Display

接下来，BIOS 通常会显示系统的一些详细信息。主要信息包括

- 处理器
- 软盘驱动器和硬盘驱动器
- 内存
- BIOS 版本和日期
- 显示器

Any special drivers, such as the ones for small computer system **interface** (**SCSI**) adapters, are loaded from the **adapter,** and the BIOS displays the information. The BIOS then looks at the sequence of storage devices identified as boot devices in the CMOS Setup.

任何特殊驱动程序都从适配器载入，BIOS 会显示出相关信息，比如小型计算机系统接口（SCSI）适配器的驱动程序就是如此。之后，BIOS 查看 CMOS 设置中的启动设备，也就是存储设备的顺序。

Key words: CMOS（互补金属氧化物半导体），interrupt handlers（中断处理程序），miniature（微型的），graphic（图形的），ROM（只读存储器），POST（加电自检程序），RAM（随机存储器），USB（通用串型总线），peripheral（外围的），component（成分），interconnect（使互相连接），PCI（互连外围设备），interface（接口），SCSI（小型计算机系统接口），adapter（适配器）

Notes

PS/2: A lot of brand machines use PS/2 interface to connect the mouse and keyboard. The PS/2 interface is different from the traditional keyboard interface in the interface appearance and pins, while in data transfer format they are the same. Now many motherboard with the PS/2 interface socket is connected with the keyboard. The keyboard can be achieved by traditional interface PS/2 interface converter connected to the motherboard PS/2 interface socket. 很多品牌机上采用 PS/2 接口来连接鼠标和键盘。PS/2 接口与传统的键盘接口在接口外形和引脚是不同的，然而在数据传送格式上是相同的。现在很多主板用 PS/2 接口插座连接键盘，传统接口的键盘可以通过 PS/2 接口转换器连接主板 PS/2 接口插座。

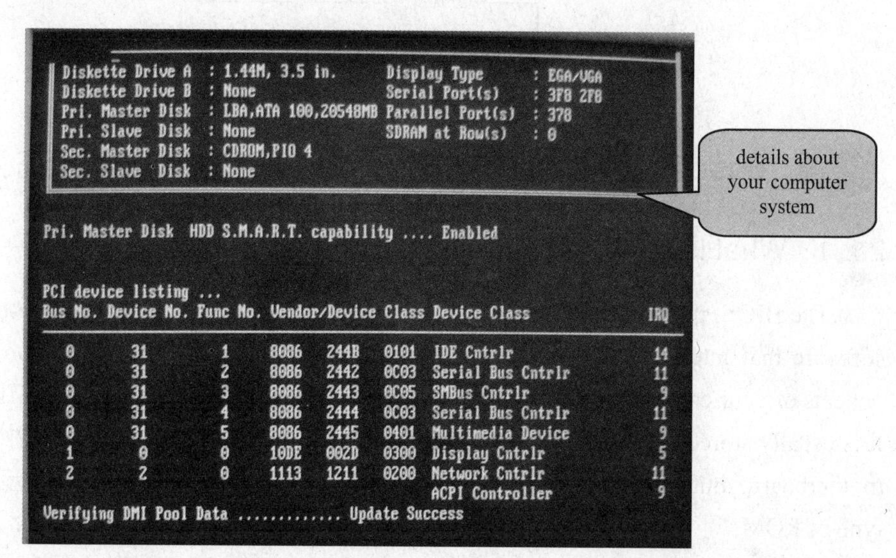

"Boot" is short for "**bootstrap**", as in an old saying, "**Lift yourself up by your bootstraps**". Boot refers to the process of launching the operating system. The BIOS will try to initiate the boot sequence from the first device. If the BIOS does not find a device, it will try the next device in the list. If it does not find the proper files on a device, the startup process will halt. If you have ever left a floppy disk in the drive when you restarted your computer, you have probably seen this message.

The BIOS has tried to boot the computer with **ignoring** the **floppy disk** left in the drive. Since it cannot find the correct system files, it can not continue. Of course, this is an easy fix. Simply take out the disk and press a key to continue.

Boot 是 Bootstrap（自举程序）的简略形式，正如古语所说，"拎着鞋带把自己提起来"。我们用 Boot 来代表启动操作系统的过程。BIOS 会从第一个启动设备开始这个启动过程。如果 BOIS 没有找到第一个启动设备，则会使用列表中的下一个设备尝试启动。如果它没有找到有关设备的正确文件，启动过程将会暂停。如果你在重新启动计算机时曾把软盘留在软盘驱动器中，你可能就见过这种暂停信息。

BIOS 试图不考虑驱动器中软盘的影响继续启动计算机，但由于它无法找到正确的系统文件，所以无法继续启动。当然，这情况很容易处理，只需取出磁盘，按下任意键就可以继续启动。

Key words: bootstrap（自举程序），Lift yourself up by your bootstraps（自举），ignore（忽视），floppy disk（软盘）

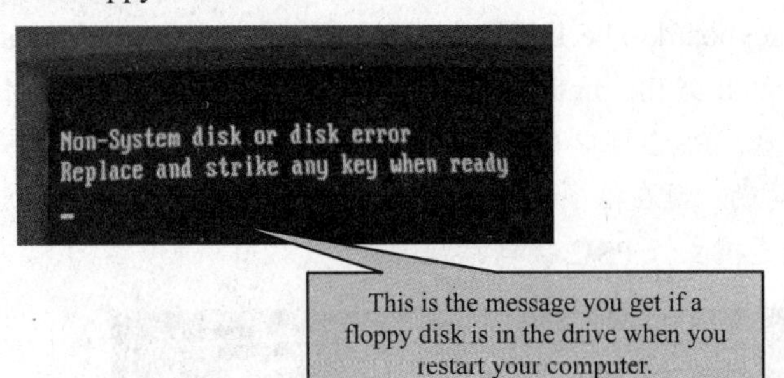

2.2 Basic Input and Output System （基本输入输出系统）

2.2.1 What Is BIOS（什么是 BIOS）

The BIOS, pronounced "bye-ose", is a special software that interfaces the major hardware components of your computer with the operating system. It is usually stored on a **Flash Memory** chip on the motherboard, but sometimes the chip is another type of ROM.

BIOS 发音为"bye-ose"，它是用来连接计算机主要硬件部分和操作系统的一种特殊软件。通常，BIOS 程序储存在主板上的一块闪存芯片中，但有时候也可能是用另一种类型的只读存储器作为其存储芯片。

On virtually every computer available, the BIOS makes sure all the other chips, hard drives, ports and **CPU** function together.

事实上，对于每台可用的计算机，BIOS 程序是用来确保所有的其他芯片、硬件驱动器、端口和 CPU 能够协同工作的。

Every desktop and laptop computer in common use today contains a **microprocessor** as its central processing unit. The microprocessor is the hardware component. To get its work done, the microprocessor executes a set of instructions known as software. You are probably very familiar with two different types of software: the operating system and the applications.

如今，每台公用台式和便携式计算机都使用微处理器作为其中央处理器。微处理器属于硬件。为了使它正常工作，微处理器需要执行一系列被称为"软件"的指令。你可能对两种类型的软件很熟悉：操作系统和应用软件。

It turns out that the BIOS is the third type of software your computer needs to operate successfully. In this part, you'll learn all about BIOS—what it does; and how to configure it.

很显然，BIOS 就是使你的计算机能够顺利工作所必需的第三种类型的软件。在这一节，你将会学习 BIOS 的相关知识——BIOS 做些什么以及如何配置它。

Inside the above-mentioned ROM three different programs are stored (actually burned). The Basic Input and Output System (BIOS) that is responsible for teaching the main processor how to deal with basic devices, such as floppy unit or hard disk and so on; the Power on Self Test (POST) which is the program in charge of the self test that is executed every time **PC** is powered up (like memory counting, for instance); and the **Setup** which is the program that allows the alteration of the parameters stored in the configuration memory (**CMOS**).

在上述只读存储器中存储着三种不同类型的程序（事实上是烧录上去的）。基本的输入输出系统（BIOS 程序）负责指导主处理器如何控制像软盘驱动器、硬盘之类的基本设备。加电自检程序（POST）是负责在每次计算机开机时执行自检测的程序（比如内存容量检测）。设置程序（Setup）则是种专门用来修改 CMOS 中存储参数的程序。

 **Key words:** Flash Memory（闪速存储器），CPU（中央处理器），microprocessor（微处理器），PC（个人计算机），Setup（设置程序），Configuration Memory（CMOS 设置存储器）

朗读音频
What Is BIOS

Motherboard ROM memory chip. BIOS, POST and Setup softwares are written in this chip.

Phoenix BIOS

AMI BIOS

2.2.2 What BIOS Does（BIOS 的任务是什么）

The BIOS software has a number of different roles, but its most important role is to load the operating system. When you turn on your computer and the microprocessor tries to execute its first instruction, it has to get that instruction from somewhere. It cannot get it from the operating system because the operating system is located on a hard disk, and the microprocessor cannot get to it without some instructions that tell it how. The BIOS provides those instructions.

BIOS 软件要执行许多不同的任务，但最重要的任务是载入操作系统。当你打开计算机时，微处理器尝试执行第一条指令，它必须从某个地方获取该指令。但由于操作系统存储在硬盘上，微处理器不可能在没有指令的情况下执行读取指令的操作，所以微处理器无法从操作系统中获得第一条指令。BIOS 就提供了那些指令。

Some of the other common tasks that the BIOS performs include:

- A power on Self Test (POST) for all of the different hardware components in the system to make sure everything is working properly.
- Activating other BIOS chips on different cards installed in the computer—For example, SCSI and graphics cards often have their own BIOS chips.
- Providing a set of low-level routines that the operating system uses to interface to different hardware devices—It is these routines that give the BIOS its name. They manage things like the keyboard, the screen, and the serial and parallel ports, especially when the computer is booting.
- Managing a collection of settings for the hard disks, clock, etc.

BOIS需要执行的其他任务如下。

- 对计算机系统中所有不同的硬件部分进行加电自检（POST），确定所有设备工作正常。
- 激活计算机中其他硬件卡上的BIOS芯片——例如，SCSI卡和图形卡通常都有自己的BIOS芯片。
- 提供一系列低级别程序连接操作系统和不同的硬件设备——BIOS正是因为这些程序而得名。这些程序可以管理各种设备，诸如键盘、屏幕、串口和并口，尤其是在计算机启动的时候。
- 管理硬盘、时钟等的设置。

When you turn on your computer, the BIOS performs the following tasks. This is its usual sequence:

1. Check the **CMOS Setup** for custom settings
2. Load the interrupt handlers and device drivers
3. Initialize registers and power management
4. Perform the Power on Self Test (POST)
5. Display system settings
6. Determine which devices are bootable
7. Initiate the bootstrap sequence

当你启动计算机时，BIOS就执行如下几项任务，并通常按下列顺序执行。

1. 检测CMOS常规设置；
2. 载入中断处理程序和设备驱动程序；
3. 初始化寄存器和电源管理；
4. 执行加电自检程序（POST）；
5. 显示系统设置；
6. 确定可启动设备；
7. 开始自举程序。

The first thing the BIOS does is checking the information stored in a tiny (64 bytes) amount of RAM located on a **complementary metal oxide semiconductor** (CMOS) chip. The CMOS Setup provides detailed information particular to your system and can be altered as your system changes. The BIOS uses this information to modify or supplement its default programming as needed.

BIOS所做的第一件事情是检查CMOS芯片上微小（64字节）的随机存储器中存储的信息。CMOS设置程序提供了关于你的计算机系统的详细信息，随着系统改变，CMOS设置也能够被修改。BIOS使用这些信息来修改或补充所需要的默认程序。

Key words: CMOS Setup（CMOS设置程序），complementary metal oxide semiconductor

（互补金属氧化物半导体）

朗读音频

What BIOS Does

2.2.3 Configuring BIOS（设置 BIOS）

In the previous list, you saw that the BIOS checks the CMOS Setup for common settings. Here's what you do to change those settings.

在前面的列表中，你可以看到 BIOS 检测 CMOS 中的设置来作为计算机的常规设置。这里我们将介绍你如何改变这些设置。

To enter the CMOS Setup, you must press a certain key or combination of keys during the initial startup sequence. Most systems use "Esc," "Del," "F1," "F2," "Ctrl-Esc" or "Ctrl-Alt-Esc" to enter setup. There is usually a line of text at the bottom of the display that tells you "Press ___ to enter setup."

要进入 CMOS 设置界面，你必须在计算机最初启动的时候按下一个特定按键或者组合按键。大多数的系统使用"Esc""Del""F1""F2""Ctrl-Esc"或者"Ctrl-Alt-Esc"按键来进入 CMOS 的设置界面。通常，屏幕下方会显示一行文字提示用户"按下___键进入设置界面"。

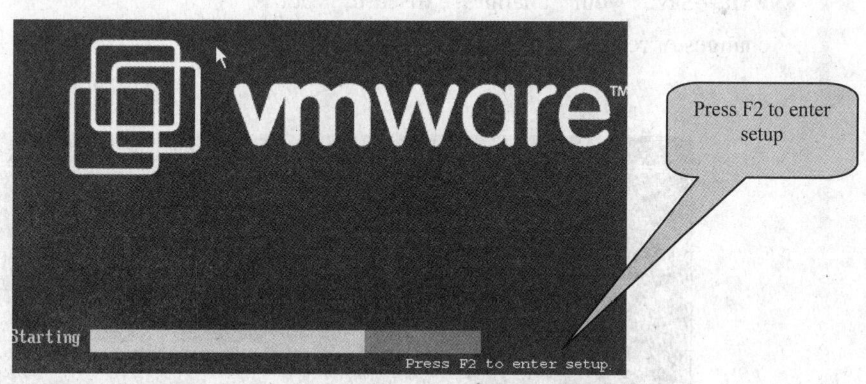

Once you have entered setup, you will see a set of text screens with a number of options. Some of these are standard, while others vary according to the BIOS **manufacturer**.

一旦你进入了设置界面，你将会看到包含一系列文字选项信息的屏幕。有些选项是标准的，而其他选项可能会根据 BIOS 制造商的不同而不同。

Common options include:

- Information—Something about the computer's information
- BIOS Version—The version number of BIOS

一般的选项都包括如下各项。

- 信息——显示关于电脑的信息；
- BIOS 版本——显示 BIOS 的版本号；

- System Time/Date—Set the system time and date
- Boot Priority—The order that BIOS will try to load the operating system
- **Plug and Play**—A standard for autodetecting connected devices should be set to “Yes” if your computer and operating system both support it
- Mouse/Keyboard—“Enable Num Lock” “Enable the Keyboard” “Auto-Detect Mouse”...
- Drive Configuration—Configure hard drives, CD-ROM and floppy drives
- **Memory**—Direct the BIOS to **shadow** to a specific memory address
- **Security**—Set a password for accessing the computer
- Power Management—Select whether to use power management, as well as set the amount of time for **standby** and **suspend**
- **Exit**—Save your changes, discard your changes or restore default settings

- 系统时间/日期——设置系统时间和日期；
- 启动顺序——设置 BIOS 载入操作系统的顺序；
- 即插即用设备——自动检测连接设备的标准，如果计算机和操作系统都支持自动检测连接的设备，应该将其设置为“Yes”；
- 鼠标/键盘——“启用数字键”“启用键盘”“自动检测鼠标”……
- 驱动器设置——设置硬盘驱动器、只读光盘和软盘驱动器；
- 内存——可命令 BIOS 将特定的内存地址指定为影子内存使用；
- 安全性——设置访问计算机的密码；
- 电源管理——选择是否使用电源管理，设置待命和挂机的时间；
- 退出——保存用户设置，放弃用户设置或者恢复默认设置。

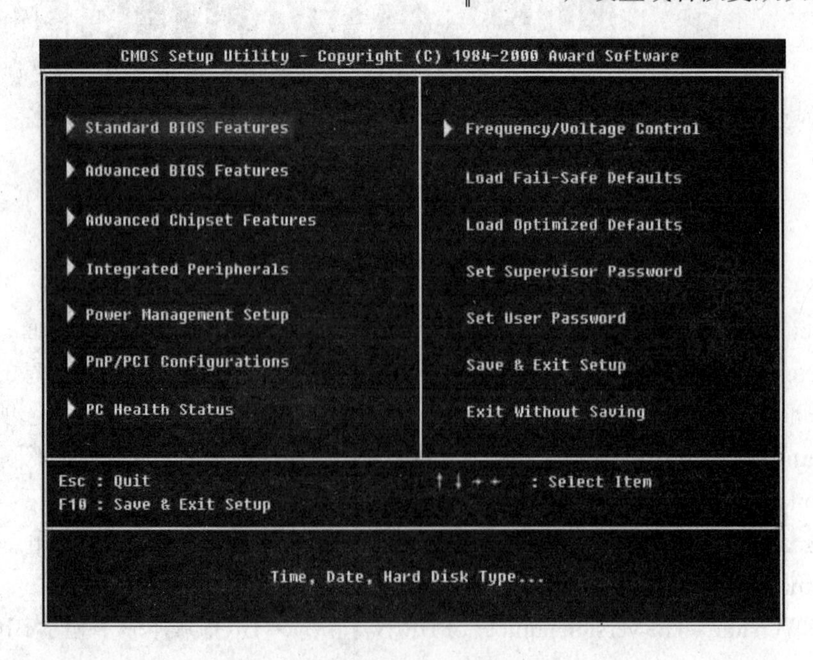

Award CMOS 设置界面

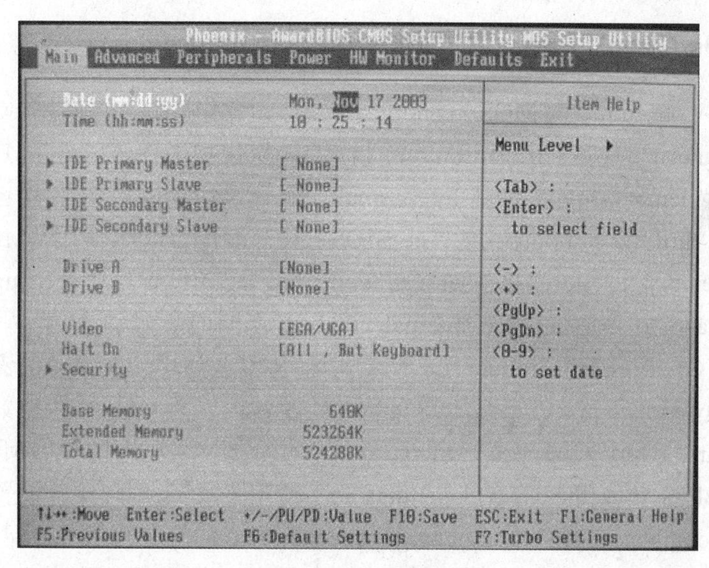

Phoenix CMOS 设置界面

Be very careful when making changes to setup. Incorrect settings may keep your computer from booting. When you finish with your changes, you should choose “Save Changes” and exit. The BIOS will then restart your computer so that the new settings take effect.

当改变 CMOS 设置时要很小心，错误的设置可能使你的计算机无法启动。当你完成了更改设置，你应该选择“保存设置”然后退出。BIOS 将会重新启动计算机从而使新的设置起效。

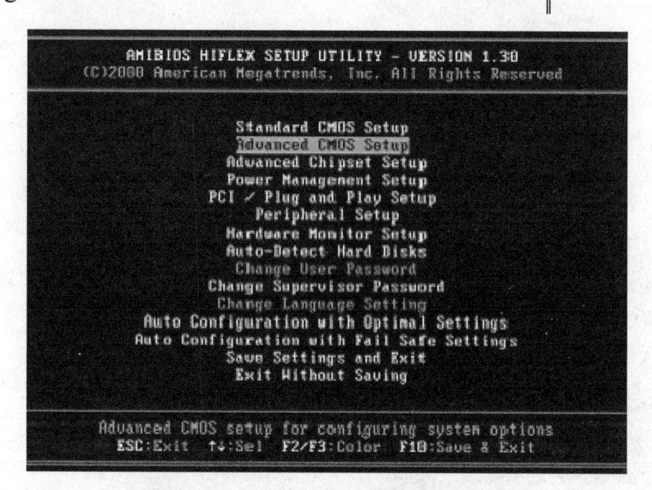

AMI CMOS 设置界面

Key words: manufacturer（制造商），Pluy and Play（即插即用），Shadow Memory（影子内存），security（安全性），standby（待机），suspend（暂停、挂起），exit（退出）

2.2.4 What Is CMOS（什么是 CMOS）

Pronounced see-moss, CMOS (Short for complementary metal oxide **semiconductor**) is a special kind of memory maintained by a small battery after you turn the computer off. The BIOS uses CMOS memory to store the settings you select in Setup and to maintain the internal real time clock (**RTC**). A small lithium or Ni-Cad battery can supply enough power to CMOS to keep the data for years. Every time you turn on your computer, the BIOS uses the CMOS settings to configure your computer. If the battery charge runs too low, the CMOS contents will be lost and POST will issue a "CMOS invalid" or "CMOS checksum invalid" message. If it happens, you may have to replace the battery. After the battery is replaced, the proper settings will need to be restored in **Setup**.

读作"see-moss"的 CMOS（互补金属氧化物半导体的缩写）是一种在计算机断电后还能够由小型电池供电的低功耗存储器。BIOS 使用 CMOS 存储器来存储用户在设置程序（Setup）中进行的设置，同时保持计算机内部的实时时钟（RTC）。使用一块很小的锂电池或镍镉电池就可以为 CMOS 供电数年之久，以保存数据。每次当用户启动计算机时，BIOS 使用 CMOS 中存储的设置来配置你的计算机。如果电池供电不足，CMOS 中的内容将会丢失，同时自检程序将会提示"CMOS 错误"或者"CMOS 校验错误"的信息。如果这种情况发生，你必须更换电池。更换电池后，需要在设置界面中重新存储相应的设置信息。

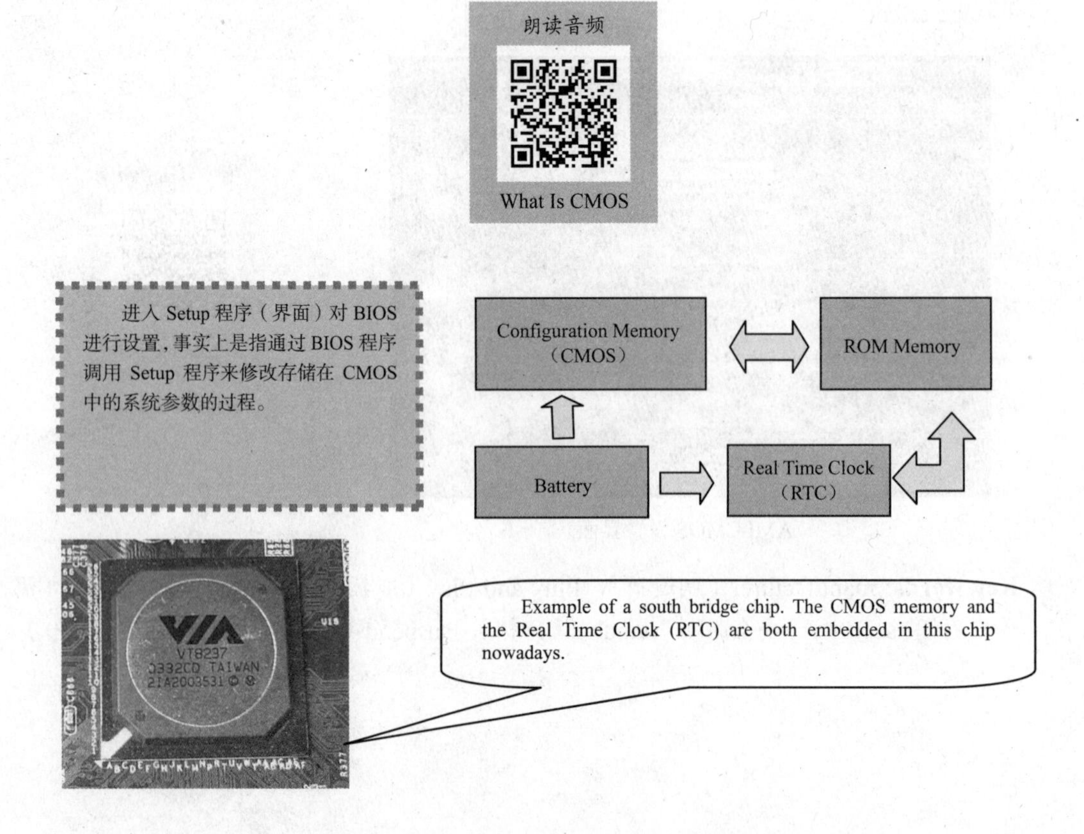

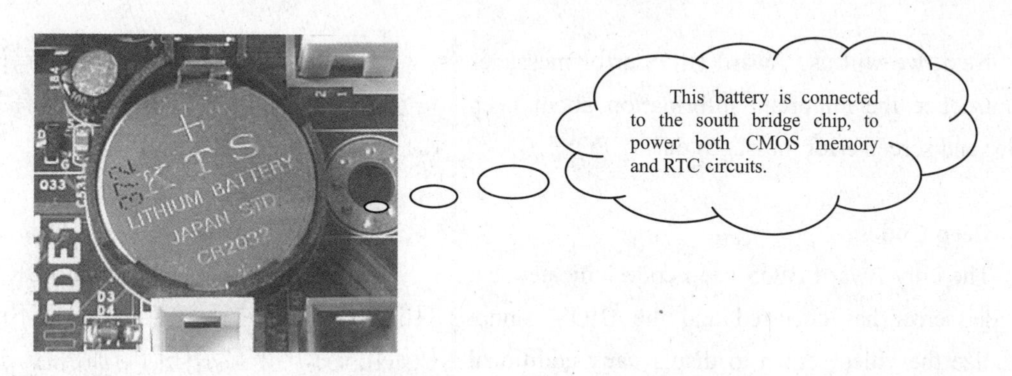

Key words: Semiconductor(半导体), RTC(实时时钟), Setup(设置程序或设置界面)

2.2.5 What Is POST(什么是 POST)

POST is an **acronym** for Power on Self Test. When a personal computer is firstly turned on, the BIOS tests and configures various components to ensure that they are operating correctly. This operation is called POST (Power on Self Test). Its meaning includes anything the BIOS does before an operating system is started.

POST 是加电自检的缩写。当个人计算机最初启动时，BIOS 检测和配置各种组件以确保它们正常工作。这个操作就叫作自检（加电自检）。自检程序包括启动操作系统之前 BIOS 所进行的所有工作。

If the BIOS detects any problems during the POST, it will attempt to continue to start the computer. However, if the problems are serious, the BIOS may be forced to halt the system.

在检测过程中，如果 BIOS 检测到任何问题，它将试图继续启动计算机。但是，如果问题严重，BIOS 可能被迫使系统暂停。

When an error is detected, the BIOS program will:

- Display the error to the screen if possible.
- Generate a **POST beep code** using the computer's internal speaker if it cannot access the display adaptor.
- Provide a **POST Code** output that can be read by using a special hardware tool.

当检测到错误时，BIOS 程序将会:

- 尽可能在屏幕上显示错误信息。
- 如果无法访问显示器，它会使用计算机内部的扬声器发出自检鸣叫声。
- 输出一个自检代码，可以通过特定的硬件工具来读取该代码。

Key words: acronym(只取首字母的缩写词), POST beep code(自检鸣叫错误), POST Code(自检代码)

Next, we will use Award BIOS error messages to introduce the important information about beep codes and screen error messages during POST.

下面，我们以 Award BIOS 的错误信息提示为例介绍在自检过程中有关鸣叫声和屏幕错误信息的重要内容。

Beep Codes

The only Award BIOS beep code indicates that a video error has occurred and the BIOS cannot initialize the video screen to display any additional information. This beep code consists of a single long beep followed by two short beeps. Any other beeps are probably a RAM (Random Access Memory) problem.

鸣叫代码

当有显示故障发生时，Award BIOS 只能发出鸣叫声提示错误信息，而不能初始化显示屏来显示任何附加信息。这种鸣叫声由一声长鸣紧接着两声短鸣构成。其他的鸣叫类型可能是提示随机存储器故障。

Screen Error Messages

The following messages are examples of messages including errors detected by the BIOS during POST and a description of what they mean and/or what you may do to correct the error.

屏幕错误信息

下面的信息示例是些在自检过程中 BIOS 检测到的错误信息及其含义和用户可以进行的相应处理措施。

BIOS ROM checksum error—System halted

The checksum of the BIOS code in the BIOS chip is incorrect, indicating the BIOS code may have become corrupt. Contact your system dealer to replace the BIOS.

BIOS 校验错误——系统暂停

BIOS 芯片中的 BIOS 代码检测错误，意味着 BIOS 代码可能已经被破坏。用户需要和计算机系统经销商联系更换 BIOS 芯片。

CMOS battery failed

The CMOS battery is no longer functional. Contact your system dealer for a battery replacement.

CMOS 电池失效

CMOS 电池不再有效。用户需要联系系统经销商更换 CMOS 电池。

CMOS checksum error—Defaults loaded

Checksum of CMOS is incorrect, so the system loads the default equipment configuration. A checksum error may indicate that CMOS has become corrupt. This error may have been caused by a weak battery. Check the battery and replace if necessary.

CMOS 校验错误——载入默认设置

CMOS 校验错误，于是系统载入默认的设备配置。校验错误可能表示 CMOS 信息被破坏。发生这种错误的原因可能是电池电量不足，如果必要的话可以先更换电池。

Display switch is set incorrectly

The display switch on the motherboard can be set to either monochrome or color. This message indicates the switch is set to a different setting than indicated in Setup. Determine which setting is correct, and then either turn off the system and change the jumper, or enter Setup and change the VIDEO selection.

显示开关设置错误

主板上的显示开关可以设置为单色或彩色。这条错误信息表示开关的设置与 CMOS 中的设置不相符。首先确定哪一个设置是正确的，然后关闭系统改变跳线或者进入 CMOS 设置界面更改“显示”选项。

Press Esc to skip memory test

The user may press Esc to skip the full memory test.

按下 Esc 键跳过内存检测

用户可以按下 ESC 键跳过完整的内存检测。

Floppy disk(s) fail

Cannot find or initialize the floppy drive controller or the drive. Make sure the controller is installed correctly. If no floppy drives are installed, be sure the Diskette Drive selection in Setup is set to NONE or AUTO.

软盘驱动器驱动失败

无法找到或初始化软盘驱动控制器或驱动器。用户需要确定控制器安装是否正确。如果没有安装软盘驱动器，则要将设置界面中的“磁盘驱动器选择”选项设置为“无”或是“自动”。

Hard disk initializing—Please wait a moment

Some hard drives require extra time to initialize.

硬盘正在初始化——请稍候

有些硬盘启动器需要额外的时间来初始化。

Hard disk install failures

Cannot find or initialize the hard drive controller or the drive. Make sure the controller is installed correctly. If no hard drives are installed, be sure the Hard Drive selection in Setup is set to NONE.

硬盘安装失败

无法找到或初始化硬盘驱动控制器或驱动器。用户需要确定控制器安装是否正确。如果没有安装硬盘驱动器，则要将设置界面中的“硬盘选择”选项设置为“无”。

Hard disk(s) diagnosis fail

The system may run specific disk diagnostic routines. This message appears if one or more hard disks return to an error when the diagnostics run.

执行硬盘诊断时发生错误

系统可以运行特定的硬盘诊断程序。当诊断程序运行时，如果一个或多个硬盘返回错误，屏幕上就会显示这条信息。

Keyboard error or no keyboard present

Cannot initialize the keyboard. Make sure the keyboard is attached correctly and no keys are pressed during POST. To purposely configure the system without a keyboard, set "the error halt condition" in Setup to "halt on all, but keyboard". The BIOS then ignores the missing keyboard during POST.

键盘错误或未接键盘

无法初始化键盘。确定键盘连接正确且在自检过程中没有按下过任何键。如果有意要配置没有键盘的系统，则要将设置界面中 "错误暂停条件" 选项设置为"除键盘外，所有错误暂停"。那么BIOS在自检过程中会忽略因为没有键盘而产生的错误。

Keyboard is locked out—Unlock the key

This message usually indicates that one or more keys have been pressed during the keyboard tests. Be sure no objects are resting on the keyboard.

键盘被锁定——解除键盘锁定

通常，这条信息表示在键盘检测过程中用户按下了一个或多个按键。用户需要确定没有物品按压在键盘上。

Memory test

This message displays during a full memory test, counting down the memory areas being tested.

内存检测

在进行完整内存检测时显示这条信息，以倒数方式计数被检测的内存区域。

Memory test fail

If POST detects an error during memory testing, additional information appears giving specifics about the type and location of the memory error.

内存检测失败

如果在内存检测中自检程序检测到错误则会出现这条信息，附加信息还会显示出内存错误的类型、位置的详细信息。

Override enabled—Defaults loaded

If the system cannot boot using the current CMOS configuration, the BIOS can override the current configuration with a set of BIOS defaults designed for the most stable, minimal-performance system operations.

当前CMOS设置无法启动系统——载入BIOS默认设置

如果系统无法使用当前的 CMOS 设置进行启动，BIOS 能够跳过当前设置而使用一系列默认配置，这些默认配置是为最稳定、最小性能系统操作而专门设置的。

Press Tab to show POST screen

System **OEM**s may replace the Phoenix Technologies' AwardBIOS POST display with their own display. Including this message in the OEM display permits the operator to switch between the OEM display and the default POST display.

按Tab键可以显示自检画面

系统的原始设备制造厂商会用自己设计的显示画面来取代 Phoenix 技术的 AwardBIOS 自检显示画面。厂商的自定义画面中会包括这条信息，以提示用户可以按Tab键切换厂商的自定义画面和BIOS默认的自检画面。

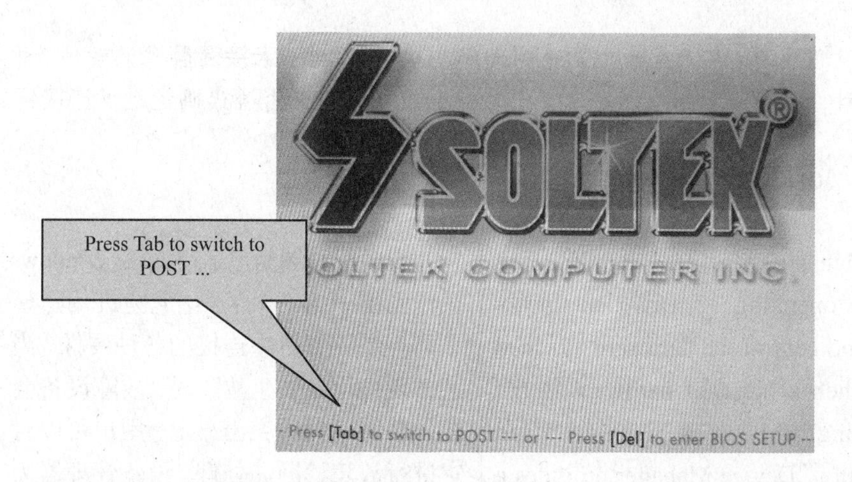

Primary master hard disk fail

POST detects an error in the primary master **IDE** hard drive.

第一 IDE 接口的主硬盘读取失败

自检程序在第一 IDE 接口主硬盘驱动器检测到错误。

Primary slave hard disk fail

POST detects an error in the secondary master IDE hard drive.

第一 IDE 接口的从硬盘读取失败

自检程序在第一 IDE 接口从硬盘驱动器检测到错误。

Secondary master hard disk fail

POST detects an error in the primary slave IDE hard drive.

第二 IDE 接口的主硬盘读取失败

自检程序在第二 IDE 接口主硬盘驱动器检测到错误。

Secondary slave hard disk fail

POST detects an error in the secondary slave IDE hard drive.

第二 IDE 接口的从硬盘读取失败

自检程序在第二 IDE 接口从硬盘驱动器检测到错误。

Key words: IDE（集成驱动器电子部件），OEM（原始设备制造厂商）

Notes

POST Codes are hexadecimal codes generated by the BIOS during the POST operation. These codes may be monitored on a special hardware “POST card” which displays them using a two digit alphanumeric display.

自检代码是在自检过程中BIOS产生的十六进制代码。这些代码可以由一种特殊的硬件“自检卡”来监测，这种“自检卡”用2个含有字母数字的代码来显示自检代码。

2.3 Device Manager（设备管理器）

The Device Manager is a tool including in Microsoft Windows operating systems that allows the user to display and control the hardware attached to the computer. When a piece of hardware is not working, the offending hardware is highlighted and the user can deal with it. Device Manager provides a graphical view of the hardwares that are installed on the computer, as well as the device drivers and resources associated with that hardware. Using Device Manager provides a central point to change the way the hardware is configured and interacts with the computer's microprocessor. In this part we will describe how to use Device Manager to manage devices in Microsoft Windows.

设备管理器是包括在微软Windows操作系统中的一个工具，它允许用户显示和控制计算机中连接的各种硬件。当某一硬件设备发生故障时，故障设备会在设备管理器中突出显示，用户可以对它进行故障诊断和处理。设备管理器为计算机中所安装的每一硬件及其设备驱动程序和资源提供了一个图形视图。使用设备管理器可以集中更改硬件配置以及硬件与计算机微处理器交互的方式。这一节中我们将描述如何利用设备管理器在微软Windows操作系统中管理各种设备。

Notes

Device Manager 1

Device Drivers: A device driver is a computer program that enables another program, typically an operating system (e.g. Windows, Linux) to interact with a hardware device. A driver is essentially an instruction manual that provides the operating system with the information on how to control and communicate with a particular piece of hardware.

设备驱动程序：设备驱动程序是一个计算机程序，它可以使另一个程序（通常指操作系统，如Windows、Linux）和硬件设备相连接。驱动程序其实是一个指令手册，它向操作系统提供了有关如何控制特定硬件设备并与之通信的相关信息。

Device Manager allows the following functions:

- Determine if the hardware on your computer is working properly.
- Change hardware configuration settings.
- Identify the device drivers that are loaded for each device and obtain information about each device driver.
- Change advanced settings and properties for devices.

设备管理器具有下列功能。

- 确定计算机上的硬件是否工作正常。
- 更改硬件配置设置。
- 认证每个设备加载的设备驱动程序，并获取每个设备驱动程序的有关信息。
- 更改设备的高级设置和属性。

- Install updated device drivers.
- Disable, enable, and uninstall devices.
- Reinstall the previous version of a driver.
- Identify device conflicts and manually configure resource settings.
- Print a summary of the devices that are installed on your computer.

Typically, Device Manager is used to check the status of computer hardware and update device drivers on the computer. If you are an advanced user, and you have a thorough understanding of computer hardware, you can use Device Manager's diagnostic features to resolve device conflicts, and change resource settings.

- 安装更新的设备驱动程序。
- 禁用、启用和卸载设备。
- 重新安装驱动程序的前一版本。
- 找出设备冲突并手动配置资源设置。
- 打印计算机上所安装设备的摘要信息。

通常，设备管理器用于检查计算机硬件的状态以及更新计算机中的设备驱动程序。如果你是高级用户并且通晓计算机硬件知识，则可以使用设备管理器的诊断功能来消除设备冲突以及更改资源设置。

Device Manager 2

2.3.1 How to Open Device Manager（如何打开设备管理器）

The Control Panel method is probably the most straightforward way to get there, but we go over all your options below.

Follow the easy steps below to open Device Manager in Windows.

Note: You can open Device Manager as described below in any version of Windows, including Windows 10, Windows 8, Windows 7, Windows Vista, and Windows XP. See what version of Windows do you have? If you're not sure.

Opening Device Manager should only take a minute or so, no matter which version of Windows you're using. See other ways to open Device Manager towards the bottom of the page for some

使用控制面板可能是最直接的方法，但我们将在下面讨论所有的选项。

按照下面的简单步骤在 Windows 中打开设备管理器。

注意：你可以在 Windows 的任何版本，如 Windows 10、Windows 8、Windows 7、Windows Vista 和 Windows XP 中打开设备管理器。如果你不确定，那要先看看你的 Windows 是什么版本。

无论你使用的是哪个版本的Windows，打开其设备管理器只需要一分钟左右。可以尝试使用其他方法打开设备管理器，至少在某些版本的 Windows 中，在页面的底

other, arguably faster, ways in at least some versions of Windows.

部有一些其他更快的方法。

How to open Device Manager via Control Panel?

如何通过控制面板打开设备管理器?

1. Open Control Panel.

1．打开控制面板。

Depending on your version of Windows, Control Panel is usually available from the Start Menu or the Apps screen.

根据 Windows 的版本，控制面板通常可以从“开始”菜单或应用程序屏幕获得。

In Windows 10 and Windows 8, assuming you're using a keyboard or mouse, the fastest way is through the Power User Menu - just press the WIN (Windows) key and the X key together.

在 Windows 10 和 Windows 8 中，假设你使用的是键盘或鼠标，最快的方法是通过“电源用户”菜单——只需一起按下 WIN（Windows）键和 X 键就可以了。

2. What you do next depends on what Windows operating system you're using:

2．下一步做什么取决于你使用的 Windows 操作系统。

In Windows 10 and Windows 8, tap or click on the Hardware and Sound link. You could also jump right to Device Manager through the Power User Menu and not have to go through Control Panel.

在 Windows 10 和 Windows 8 中，点击或单击硬件和声音链接。你也可以通过电源用户菜单直接跳转到设备管理器，而不必经过控制面板。

In Windows 7, click System and Security.

在 Windows 7 中，单击系统和安全。

In Windows Vista, choose System and Maintenance.

在 Windows Vista 中，选择系统和维护。

In Windows XP, click Performance and Maintenance.

在 Windows XP 中，单击性能和维护。

Note: If you don't see these options, your Control Panel view may be set to Large icons, Small icons, or Classic View, depending on your version of Windows. If so, find and choose Device Manager from the big collection of icons you see and then skip to Step 4 below.

注意：如果你没有看到这些选项，你的控制面板视图可能会被设置为大图标、小图标或经典视图，这取决于你的 Windows 版本。如果是的话，从你所看到的图标集合中找到并选择设备管理器，然后跳到下面的步骤 4。

3. From this Control Panel screen, look for and choose Device Manager.

3．从这个控制面板屏幕，查找并选择设备管理器。

In Windows 10 and Windows 8, check under the Devices and Printers heading. In Windows 7, look under System. In Windows Vista, you'll find Device Manager towards the bottom of the window.

在 Windows 10 和 Windows 8 中，在设备和打印机标题下进行查找。在 Windows 7 中，在系统下查找。在 Windows Vista 中，您将在窗口底部找到设备管理器。

Windows XP Only: You have a few extra steps since Device Manager isn't as easily available in your version of Windows. From the open Control Panel window, click System, choose the Hardware tab, and then click the Device Manager button.

仅在 Windows XP 中，你有一些额外的步骤需要设置，因为设备管理器在此 Windows 版本中不是那么容易找到。先打开“控制面板”窗口，单击“系统”，选择“硬件”选项卡，然后单击“设备管理器”按钮。

4. With Device Manager now open, you can view a device's status, update the device drivers, enable devices, disable devices, or do whatever other hardware management you came here to do.

4．设备管理器现在已经打开，你可以查看设备的状态，更新设备驱动程序，启用设备，禁用设备，或者做其他的硬件管理工作。

Other Ways to Open Device Manager.

打开设备管理器的其他方法。

If you're comfortable with the command-line in Windows, specifically Command Prompt, one really quick way to start Device Manager in any version of Windows is via its run command, devmgmt.msc.

如果你熟悉 Windows 的命令行，尤其是命令提示符的话，那么通过运行命令 devmgmt.msc，是打开任何版本 Windows 设备管理器的非常快速的方法。

See How to Access Device Manager From the Command Prompt for a full walkthrough, including a few other commands that work too.

请参见如何从命令提示符访问设备管理器以进行全面演练，包括其他一些工作的命令。

The command-line method really comes in handy when you need to bring up Device Manager but your mouse won't work or your computer is having a problem that prevents you from using it normally.

当你需要启动设备管理器，但你的鼠标不能工作或你的计算机有一个妨碍你正常使用的问题时，命令行方法真的就派上用场了。

While you probably won't ever need to open Device Manager this way, you should know that it's also available in all versions of Windows via Computer Management, part of the suite of built-in utilities called Administrative Tools.

虽然你可能不需要以这种方式打开设备管理器，但你应该知道它可以通过计算机管理在 Windows 的所有版本中获得，这是一组被称为管理工具的内置实用程序的一部分。

Device Manager takes on a slightly different look in Computer Management. Just tap or click on it from the left-margin and then use it as an integrated feature of the utility on the right.

在计算机管理中，设备管理器的外观稍有不同。只要从左边空白处点击或单击它，然后将其作为实用工具的一个集成功能就可以了。

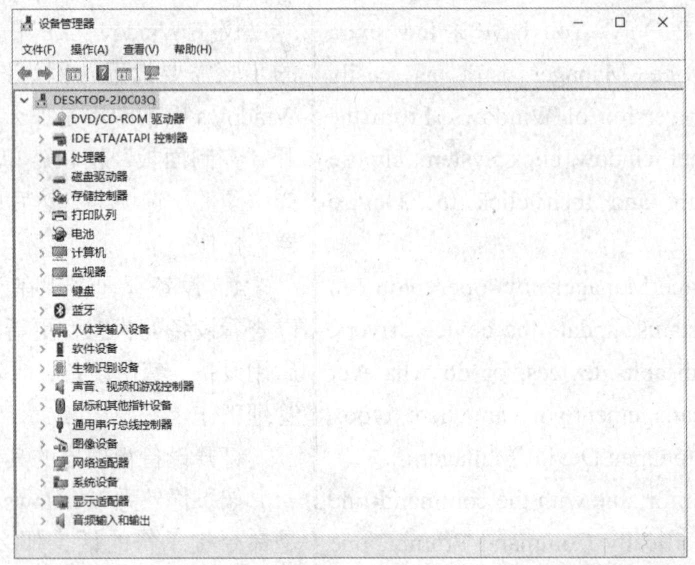

Device Manager for Windows 10

2.3.2 Viewing Information about a Device Driver（查看有关设备驱动程序的信息）

To get information about the driver for a device, perform the following steps:

1. Double-click the type of device you want to view in device manager.

2. Right-click the specific device and then click Properties.

3. On the Driver tab, click Driver Details.

Information about the Device Driver helps determine the file version of the device driver. An icon appears next to device drivers that are digitally signed.

Depending on how your computer is configured, Windows either ignores device drivers that are not digitally signed, displays a warning when it detects device drivers that are not digitally signed (the default behavior), or prevents you from installing device drivers without digital signatures.

若要获取某个设备的驱动程序的有关信息，请按照下列步骤操作。

1. 在设备管理器中双击要查看的设备的类型。

2. 右键单击特定的设备，然后单击“属性”。

3. 在驱动程序选项卡上，单击“驱动程序详细信息”。

有关设备驱动程序的信息可帮助确定设备驱动程序的文件版本。如果设备驱动程序进行了数字签名，在驱动程序的旁边就会出现一个图标。

根据计算机的配置情况，Windows 忽略未经数字签名的设备驱动程序，在检测到无数字签名的设备驱动程序时显示警告消息（默认行为），或者禁止安装无数字签名的设备驱动程序。

Notes

Microsoft 对 Windows 设备驱动程序和操作系统文件都进行了数字签名，以保证它们的质量。Microsoft 的数字签名保证：某个文件来自其制造商，并且此文件未被其他程序的安装过程所修改或改写。

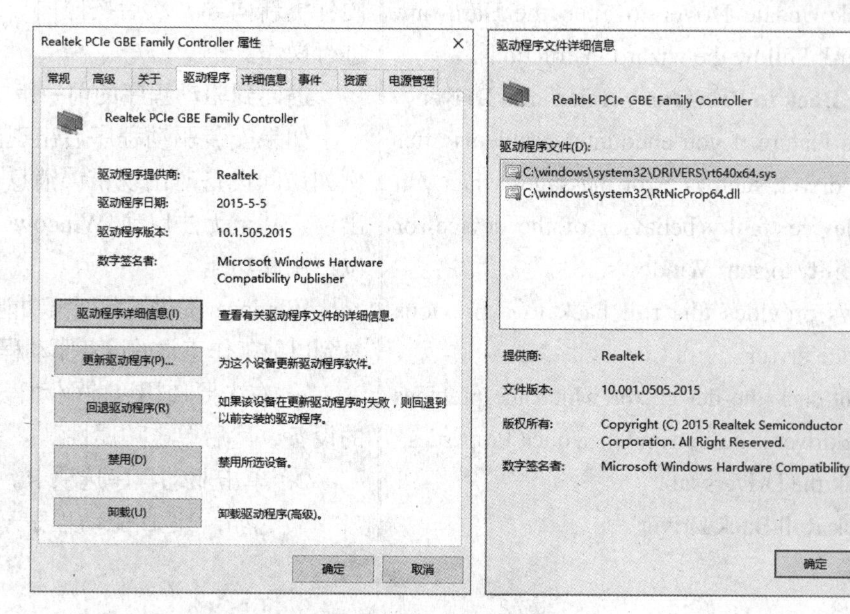

Details about a Device Driver

2.3.3 Updating or Changing a Device Driver（更新或更改设备驱动程序）

Ensure that the latest device driver for each of the devices is loaded in Windows. Manufacturers frequently update their drivers to fix problems and take advantage of operating system features. These drivers are usually available from the manufacturer's web site, and Microsoft also maintains driver files for many devices on its web sites.

Upon downloading drivers, read the manufacturer's instructions before attempting to use the files to update the device. Often, downloaded driver files are compressed into a self-executing file that needs to be extracted in order to use the driver. In the Hardware Update Wizard, click to select the Hard Disk option, then click the Browse button to locate the driver files.

确保在 Windows 中给每个设备装载了最新的驱动程序。制造商经常更新他们的驱动程序，以便纠正出现的问题并利用操作系统的功能。通常，可从制造商的网站上下载这些驱动程序，而且微软也在其网站上维护许多设备的驱动程序文件。

下载这些驱动程序后，在使用这些文件更新设备之前，请先阅读制造商的说明。通常，下载的驱动程序被压缩成一个可自执行的文件。要使用驱动程序，需要解压缩此文件。在硬件更新向导中，单击“从磁盘安装”选项，然后单击“浏览”按钮找到驱动程序文件。

How To Install a New Driver from Device Manager

1. Double-click the type of device you want to update or change.

2. Right-click the specific device driver you want to update or change.

3. Click Update Driver to open the Hardware Update Wizard. Follow the wizard instructions.

Rolling Back to Previous Version of a Driver

Use this feature if you encounter problems after you install a driver, such as error messages when you access the device, faulty behavior of the device, or even the inability to start Windows.

Windows provides this roll back to a previous working device driver:

1. Right-click the device for which the previous version of the driver is desired and then click Properties.

2. Click the Drivers tab.

3. Click Roll Back Driver.

如何从设备管理器安装新的驱动程序

1．双击要更新或更改的设备类型。

2．右击要更新或更改的特定设备驱动程序。

3．单击“更新驱动程序”以打开硬件更新向导，然后按照向导的指示进行操作。

退回到驱动程序的前一版本

如果安装驱动程序后出现问题，例如访问设备时出现错误消息，设备出错，甚至无法启动 Windows，则可以使用此功能。

Windows 提供了此回滚功能，以恢复到以前工作正常的设备驱动程序：

1．右击要使用以前驱动程序版本的设备，然后单击“属性”。

2．单击驱动程序选项卡。

3．单击“返回驱动程序”。

Updating or Changing a Device Driver

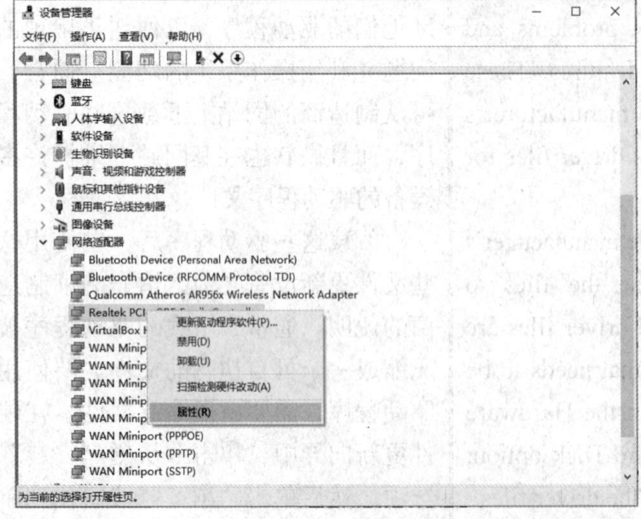

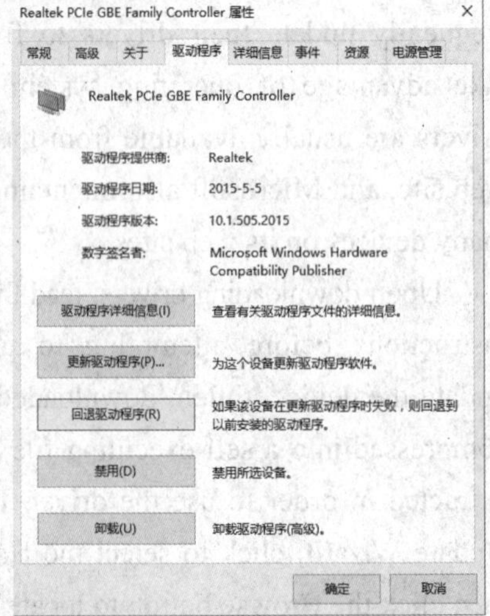

Rolling Back to Previous Version of a Driver

2.3.4 To Access the Status Information of the Specific Device(查看特定设备状态信息)

To access the status information of the device, perform the following steps:

1. Open device manager.
2. Double-click the type of device you want to access.
3. Right-click the specific device you want to access, click Properties.
4. Status information is displayed in the device status box on the General tab.

If the status is Disabled, that is usually the result of user action, and does not necessarily mean that the device has a problem. However, sometimes users disable a device because it was causing a problem, and you should try enabling it to see if it impacts another device negatively.

If the device is experiencing a problem, the Device Status box displays the type of problem. You may see a problem code, or number (or both) and a suggested solution. If you call a support line, this number can be useful for determining and diagnosing the problem.

如需查看特定设备状态，请执行以下操作。

1．打开设备管理器。

2．双击你所希望查看的设备类型。

3．右击待查设备，在快捷菜单中单击“属性”。

4．在通用选项卡上，设备状态区域内所显示的即为该设备的状态描述信息。

如果状态信息是“禁用”，那并不一定意味着设备有故障，通常是用户设置的结果。有时因为设备有问题，用户可以禁用该设备。如果遇到这种情况，你应该先启用该设备，确定它是否和其他设备冲突。

如果某种设备出现问题，问题类型将在设备状态区域内被显示出来。同时，你可能还将看到问题代码、编码及建议解决措施。如果你拨打技术支持电话，问题编码将有助于问题的确定与诊断。

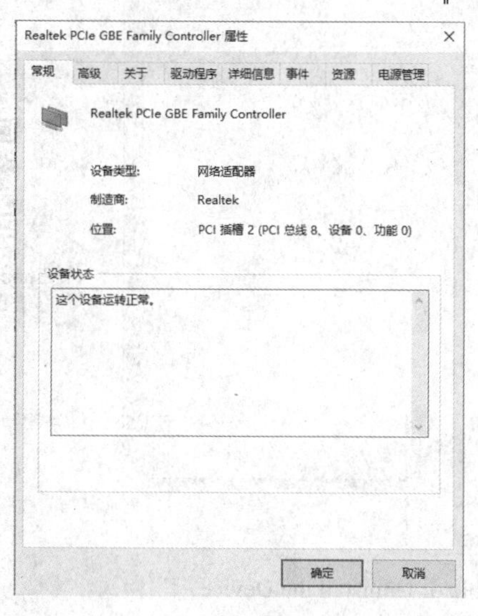

Status Information of the Specific Device

2.3.5 Enable or Uninstall the Device（启用或卸载设备）

To enable the device, please perform the following steps:

1. Open the device manager.

2. Double-click the type of device you want to access.

3. Right-click the specific device you want to access, click Enable.

To uninstall the device, please perform the following steps:

1. Open the device manager.

2. Double-click the type of device you want to access.

3. Right-click the specific device you want to access, click Uninstall.

4. Click OK button to confirm device removal.

After you finish the above-mentioned steps, turn your computer off and then take the device down.

如需启用某种设备，请执行以下步骤。

1. 打开设备管理器。

2. 双击你所希望查看的设备类型。

3. 右击你希望启用的设备，单击“启用”。

如需卸载某种设备，请执行以下操作步骤。

1. 打开设备管理器。

2. 双击你所希望查看的设备类型。

3. 右击你希望操作的设备，单击“卸载”。

4. 在确认设备删除对话框中，单击“确定”按钮。

当你完成设备卸载操作后，请关闭计算机，并从计算机上拆卸相应设备。

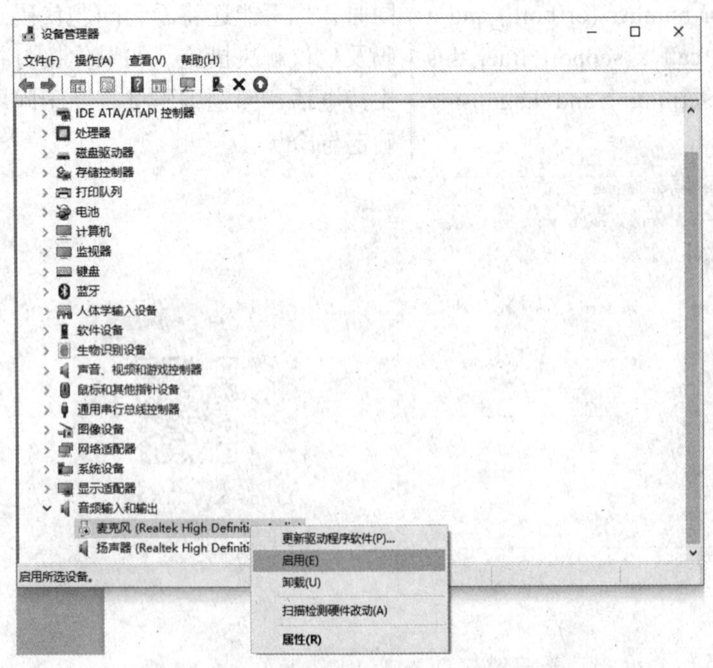

Enable or Uninstall the Device

2.4 Common DOS Commands (DOS 常用命令)

According to the above chapter, we have known something about DOS that was the first widely-installed operating system on personal computers. As we know PC-DOS and MS-DOS are almost identical and most users have referred to either of them as just "DOS". Although it is widely believed that MS-DOS is an antiquated and useless operating system with few features and capabilities, this is far from correct. In fact, although not generally publicized, MS-DOS is still used today by numerous businesses and individuals around the world. DOS was, and still is, a non-graphical line-oriented command-driven operating system, with a relatively simple interface when comparing with a GUI such as Microsoft's Windows or Apple's Macintosh.

通过前面章节的学习，我们已对DOS有所了解，DOS是第一个广泛安装于个人计算机上的操作系统。众所周知，PC-DOS与MS-DOS几乎等同，因此许多用户把它们统称为"DOS"。虽然人们大都认为MS-DOS是一个过时无用而且性能少得可怜的操作系统，但这完全是错误的观点。实际上，虽未被普及，但今天MS-DOS仍然被许多商家和个人所使用。DOS过去和现在都一直是非图形的、面向行的、命令驱动的操作系统，和诸如微软Windows或苹果Macintosh一类的图形用户接口比起来，其界面相对简单。

Before you learn some common DOS commands, make sure that you have learned how to turn on the PCs as they are configured in your classroom, that you have familiarized yourself with their keyboards and that you understand the meaning of the concept of a file and the elements of a file specification (drive, filename, and extension). Most important thing is that you can start MS-DOS in two ways. One is that you first turn on your computer, you will see the DOS command prompt after some cryptic information stops scrolling, this is the real DOS environment. And another is that enter Windows START menu and then click RUN submenu, and then input CMD.EXE or COMMAND.COM into the following message box (see Figure), at last click OK, that is the MS-DOS Windows environment.

在学习常用DOS命令之前，你要确定已经会启动教室里配置好的PC，确定你已熟知键盘，并已理解文件及其基本概念（驱动器、文件名和扩展名）的含义。关键的是能够用两种方法启动MS-DOS。一种方法是，当启动计算机后，便可在一些神秘的开机信息结束停止滚屏时看到DOS命令提示符，这是真正的DOS环境。另一种方法是，进入Windows"开始"菜单，单击"运行"子菜单，然后在如图所示的消息框中输入"CMD.EXE"或"COMMAND.COM"，最后点击"确定"，便进入了MS-DOS的窗口环境。

After these preparations, let's begin our big project. You will quickly find that the best way to learn how to use a computer is through experimentation. That is, once you have learned a command, try some variations until they don't work.

经过以上准备，我们可以进入正常的学习计划了。你将迅速发现，学习如何使用计算机的最佳途径是做实验。也就是说，一旦学了某条命令，可以尝试一些变化，直到不能再进行变通为止。

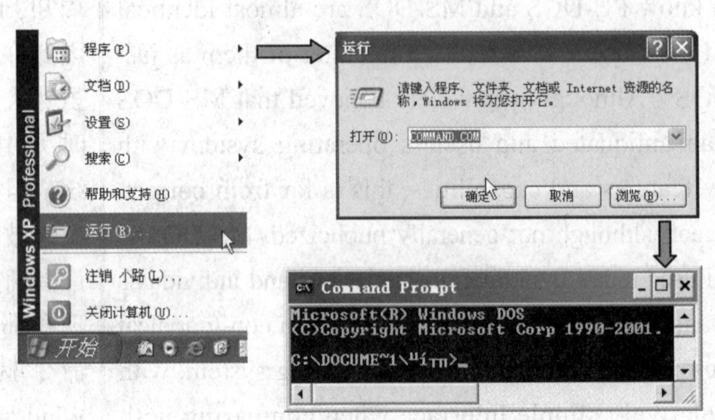

Starting DOS in Windows Environment

2.4.1 dir（文件列表命令）

Purpose	The DIRECTORY command lists the names and sizes of all files located on a particular disk （列出指定路径下所有文件清单）	
Syntax	DIR [drive:][path][filename] [/A[[:]attributes]] [/B] [/C] [/D] [/L] [/N] [/O[[:]sortorder]] [/P] [/Q] [/S] [/T[[:]timefield]] [/W] [/X] [/4]	
Examples	C:\>dir	Lists all files and directories in the directory that you are currently in (drive C) 列出当前目录中的所有文件和目录（驱动器 C）
	C:\>dir D:	Shows directory of drive D 显示驱动器 D 目录
	C:\>dir/w	If you don't need the information on the date/time and other information on the files, you can use this command to list just the files and directories going horizontally, taking as little as space needed 如果您不需要关于日期/时间的信息和其他关于文件的信息，可以使用此命令以水平的方式列出文件和目录及所占用的空间
	C:\>dir/p	If the directory has a lot of files, and you cannot read all the files as they scroll by, you can use this command, and it will display all files one page at a time 如果目录中有很多文件，当它们滚动时你不能看清楚所有文件，你就可以使用这个命令，它会一页一页地显示出所有文件

续表

Try It Now!	*Try this command 'C:\>dir/w D', the result is...* *请尝试此命令"C:\>dir/w D",结果是……* *Try the command 'C:\>dir/s/p/w', you will get what* *请尝试此命令 "C:\>dir/s/p/w",你会得到什么*
Tips	①All the files are listed at the screen, you can stop the display by typing Ctrl+C 所有的文件都列在屏幕上,你可以通过键入 Ctrl+C 停止显示 ②Two little characters called wildcard, '*' and '?', will make your life with computers much easier. E.g. C:\>dir a:*.exe, lists all files on the A drive with an extension of 'EXE' "*"和"?"这两个小字符叫做通配符,它们会让你使用计算机变得更加容易。例如,命令 C:\>dir a:*.exe,是在一个驱动器上列出一个扩展名为 exe 的所有文件

2.4.2 type(显示文件内容)

Purpose	The TYPE command is to display the contents of a file TYPE 命令是显示文件的内容	
Syntax	TYPE [drive:][path][filename]	
Example	D:\>type C:\autoexec.bat	Display the contents of the file autoexec.bat on drive C 显示驱动器 C 的 AUTOEXEC.BAT 文件的内容
Tips	If you display files that contain special (non-text) characters, these characters may have unpredictable effects on your display. 如果显示包含特殊字符(非文本)的文件,这些字符可能出现不可预知的显示效果 Wildcard characters cannot be used with this command in either the filename or the extension. 通配符不能在文件名或扩展名中使用	

2.4.3 copy(复制命令)

Purpose	The COPY command is usually used to copy one or more files from one location to another. However, COPY can also be used to create new files 文件复制命令用来把一个或多个文件从一个位置复制至另一位置,还可用作创建新文件。	
Syntax	COPY [/A | /B] source [/A | /B] [+ source [/A | /B] [+ ...]] [destination] [/A | /B]] [/V] [/Y | /-Y]	
Examples	C:\>copy test.doc D:	Copy the file TEST.DOC from drive C to drive D (with the same name). 从 C 盘到 D 盘复制文件 TEST.DOC(同名)
	C:\>copy test.doc test2	Copy the file TEST.DOC to the current directory with the new name, TEST2. 在当前目录下将文件 TEST.DOC 复制为新文件名,TEST2

续表

Examples	C:\>copy test1.doc+test2.doc :test3	Copy and combine (concatenate) the files TEST1.DOC and TEST2.DOC to a new file, TEST3 将文件TEST1.DOC和TEST2.DOC复制且合并到一个新的文件，TEST3
Try It Now!	*Try this command 'C:\> copy command.* D:', the result is...* *试试这个命令"C:\> copy command.* D:"，结果是……*	
	Try the command 'C:\> copy a:mymap.dwg c:\maps', you will get what *试着使用命令"C:\> copy a:mymap.dwg c:\maps"，你会得到什么*	
Tips	①Files can be copied to the same directory only if they are copied with a new name 只有在以新名称复制时，才能将文件复制到同一目录中 ②Unlike the BACKUP command, copied files are stored in the same format they are found in. The copied files can be used just as you would use the original one (whether the copied file is a data file or a program) 与备份命令不同，复制文件的存储格式与它们所创建的格式相同。复制的文件可以像原始文件一样使用（无论复制的文件是数据文件还是程序）	

2.4.4 ren（文件改名命令）

Purpose	The RENAME command permits users to change the name of a file without making a copy of it 改名命令允许用户修改文件名而不是复制这个文件	
Syntax	RENAME/REN [drive:][path][directoryname1 | filename1] [directoryname2 | filename2]	
Examples	C:\>ren a:sales.txt newsales.txt	Change the name of the file SALES.TXT on drive A to NEWSALES.TXT on drive A 在驱动器A上将文件SALES.TXT的名称改为NEWSALES.TXT
	C:\>rename *.txt *.bak	Rename all text files to files with .bak extension 将所有文本文件重命名为带有扩展名.bak的文件
Tip	The file name and extension must be complete for the source file, and no drive specification is given for the target. Because renaming can only occur on a single disk drive (otherwise COPY must be used) 源文件的文件名和扩展名必须是完整的，并且对目标文件没有特别的驱动器说明。因为重命名只能发生在单个磁盘驱动器上（否则必须使用副本）	

2.4.5 del（文件删除命令）

Purpose	The ERASE/DEL command deletes (erases) one or more files from disk 删除命令可将磁盘上的一个或多个文件删除
Syntax	ERASE/DEL [/P] [/F] [/S] [/Q] [/A[[:]attributes]] filenames

续表

Examples	C:\>erase a:myfile.txt	Erase the file MYFILE.TXT from the diskette in the A drive. If no drive specification is entered, the system looks to delete the specified file from drive C (in this case) 在驱动器 A 磁盘上删除文件 MYFILE.TXT。如果没有输入驱动器的说明，系统将在驱动器 C 上删除指定的文件（在本例中）
	C:\>del *.*	Erase all the files in the current directory. If you use this form of the command (to delete all files in a directory), the program will display the prompt 'Are you sure (Y/N)?' If you are sure you are deleting the files you want to delete, press the letter Y key to start the erasing process 清除当前目录中的所有文件。如果你使用此命令的形式（删除目录中的所有文件），程序将显示提示符"你确定吗？（Y / N）"如果你确定要删除想要删除的文件，请按字母 Y 键开始清除过程
Tip	Users will not receive a confirmation prompt when performing this command. Once executed, all files will be deleted unless the files have a read-only attribute and the /F switch is not being used. 执行此命令时，用户将不会收到确认提示。一旦执行，所有文件将被删除，除非文件具有只读属性，并且不使用/F 开关	

2.4.6 format（磁盘格式化命令）

Purpose	The FORMAT command checks a diskette for flaws and creates a directory where all the names of the diskette's files will be stored 磁盘格式化命令检查磁盘碎片，并且建立一个可保存所有磁盘文件名称的目录	
Syntax	FORMAT volume [/FS:file-system] [/V:label] [/Q] [/A:size] [/C] [/X] FORMAT volume [/Q] [/1] [/4] [/8]	
Examples	C:\>format a:	This form would erase all the contents off a disk. Commonly used on a diskette that has not been formatted or on a diskette you wish to erase. When the FORMAT operation is complete, the system will ask if you wish to FORMAT more diskettes. If you are working with only one diskette, answer N and carry on with your work. If you wish to FORMAT several diskettes, answer Y until you have finished formatting all your diskettes 这个命令将删除磁盘上的所有内容。通常用于未格式化的磁盘或希望擦除的磁盘上。当格式化操作完成后，系统会询问你是否希望格式化更多的磁盘。如果你只使用一个磁盘，回答 N，并继续你的工作。如果你想格式化多个磁盘，回答 Y，直到你将所有的磁盘完成格式化
	C:\> format a: /q	Quickly erases all the contents of a floppy diskette. Commonly used to quickly erase all information on the diskette 快速清除磁盘所有的内容。通常用于快速擦除磁盘上的所有信息
Tip	Executing the format command with a diskette which already contains files will result in the deletion of all the contents of the entire disk. That is to say, if you format an old diskette make sure it contains nothing you wish to save 针对包含文件的磁盘执行格式化命令将导致删除整个磁盘的所有内容。也就是说，如果格式化旧磁盘，请确保它不包含您希望保存的任何内容	

2.4.7 Common Directory Operating Commands（常见目录操作命令）

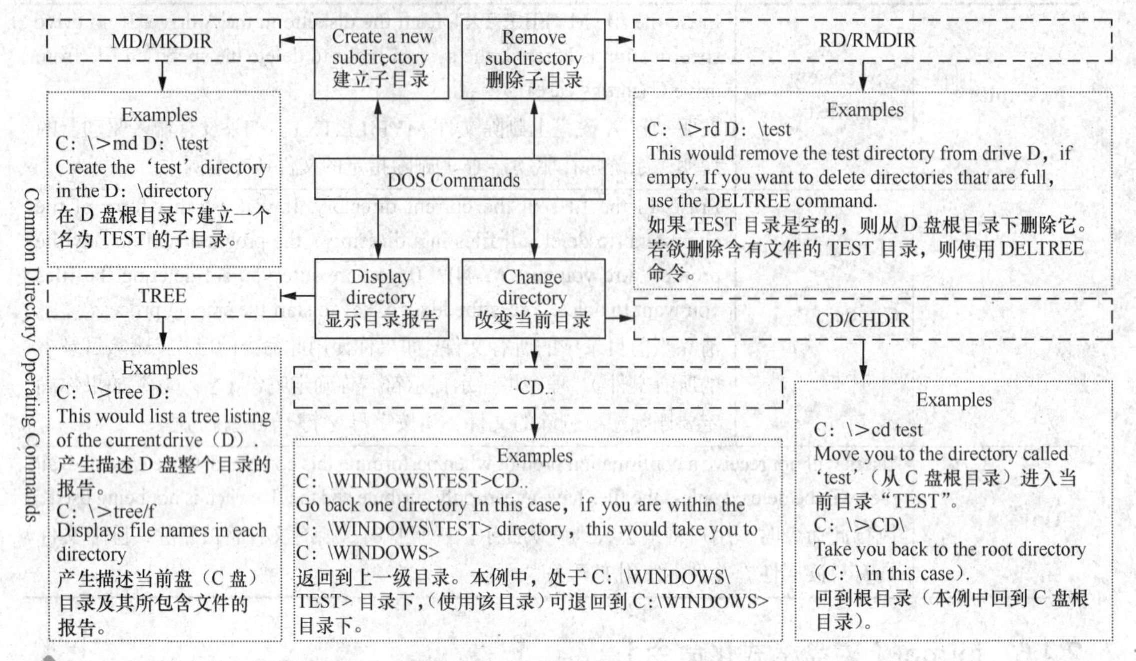

Key words: DIR（查看目录命令），TYPE（显示文件内容命令），COPY（文件复制命令），REN（RENAME，改名命令），ERASE/DEL（删除命令），FORMAT（磁盘格式化命令），MD/MKDIR（建立子目录命令），RD/RMDIR（删除子目录命令），TREE（显示目录报告命令），CD/CHDIR/CD（改变当前目录命令）

2.5 Situational Dialogue（情境对话）

Setting Passwords

Peter is a freshman who's just bought a computer for one year. Certainly, he is a newbie. Seiya is Peter's good friend who majors in computer science at college.

Peter: I can't stand my little brother. Er.... He's always making some trouble with my lovely computer. I will set a password for it so he couldn't touch it! But how? Please!

Seiya: Just take it easy! Let me show you my magic box called BIOS Settings—Security/Password Settings. You know what I mean?

设置密码

皮特是大一新生，计算机刚买了一年，自然也是计算机新手。

赛亚是皮特的好朋友，在大学主修计算机科学。

皮特：我再也不能容忍我的小弟了。嗯……他总是胡乱地使用我心爱的计算机。我要为计算机设个密码，这样他就甭想再碰我的小宝贝了！但是如何设置呢？帮我啊！

赛亚：别紧张！让我给你看看一个叫 BIOS 的设置——安全/密码设置的神奇宝盒。你知道我的意思吗？

Peter: Ah, yeah! Set security passwords to control access to the system at boot time and/or when entering the BIOS setup program. Some systems have a single password, while many new ones may have two: a supervisor and a user password. Mine has two, hasn't it?

Seiya: Not bad! But you should write it down at least two places when setting the password. If you lose the password you will be locked out of your own system, and then you will have to resort to techniques like these, for example, using password cracking software, resetting the CMOS, using the jumpers, removing the CMOS battery for at least 10 minutes and so on. For most users, a password is unnecessary while in a shared office environment such as in your room, it can be very helpful.

Peter: Sound like a specialist, I'll be careful. And I wonder what's the difference between a supervisor and a common user password?

Seiya: Good question! Generally speaking, the supervisor password is the higher level password of the present system, and the user password is the lower one. When you set user password in BIOS, you can only look at setting information. When you set supervisor password in BIOS, you can change it after verification .

Peter: I see, anything else?

Seiya: Well, you should notice the BIOS passwords are not for entering Windows or other operating system. BIOS passwords are for setting BIOS parameters, or for booting the DOS.

Peter: Wonderful. And the first step is to enter CMOS setup mode, choose supervisor password or user password menu, and then enter the code, right?

皮特：啊，知道！就是在系统引导和（或）欲进入 BIOS 设置程序时，设置控制访问的安全密码。有的系统只有一个密码，现在许多新系统则可设置两个密码：超级用户密码和普通用户密码。我的计算机有两个，对吧？

赛亚：你还可以啊！但在设置密码时必须强调的是把密码分别放置于至少两个地方。如果密码丢失那么你的系统会被封锁起来，那就得采取一些措施了，诸如，使用密码破解软件，使用跳线重置 CMOS，移出 CMOS 中的电池至少 10 分钟左右等。对于绝大多数用户来讲无需设置密码。但在公共的办公环境下（如在你的办公室），它却是十分有用的。

皮特：听起来你很在行啊！我会注意的。我想知道超级用户密码和普通用户密码有何不同？

赛亚：问得好！简单说，超级用户密码是现行系统下的高级别密码而普通用户密码则是低级别密码。在 BIOS 中设置了普通用户密码，任何人进入 BIOS 中，只能查看设置信息。而设置了超级用户密码，通过验证进入 BIOS 中，则可以修改设置信息。

皮特：明白了！还有要注意的吗？

赛亚：还有，要注意我们说的 BIOS 密码不是进入 Windows 或其他操作系统的密码。BIOS 密码是为设置 BIOS 参数或引导 DOS 而设的。

皮特：好的。第一步就是进入 CMOS 设置模式，选择超级用户密码或普通用户密码菜单，输入密码，是吧？

Seiya: Not quite right. Make sure that you input the same code at two times, I mean, at the first time and the confirmed time. This is one thing you must notice. The second is about the Security Option at Sub Menu of BIOS Feature Setup.

Peter: So what?

Seiya: This option has two choice menus, one is System—the system will not boot and accessing to Setup will be denied if the correct password is not entered at the prompt, and the other is Setup—the system will boot, but accessing to Setup will be denied if the correct password is not entered at the prompt.

Peter: Oh, I've learned so much today, I can't wait to try it now! Would you like to help me?

Seiya: No problem!

赛亚：不完全对。确保两次输入的密码一致，我指的是第一次输入以及密码确认，这是需要注意的一个问题。第二个问题则是 BIOS 特性设置菜单下的安全选项。

皮特：怎么样？

赛亚：该选项有两个选择：一是系统选项，若未在提示符下输入正确的密码，则系统不会引导并进入设置模式；二是设置选项，此时如果也未输入正确的密码，系统可以被引导，但不能进入设置模式。

皮特：哦，今天学太多知识了！我现在已迫不及待地想试试了。你愿意帮我吗？

赛亚：没问题！

2.6 Reading and Compacting（对照阅读）

Robots Technology

Today, even schoolchildren know the word "robot" and what it means because they have seen so many robots, real and imaginary, in exhibition halls, books, cartoons, and films. In some developed countries, such as Japan and the United States, industrial robots of various types are working in workshops day and night, performing repetitive, monotonous, simple and delicate jobs, such as assembling, painting, welding, cleaning, conveying, stacking, inspecting, ect. These robots deserve the nickname "blue-collar workers" as they are computerized, reprogrammable, automatic, multi-functional machine that can take over those jobs once considered exclusive for human workers, and perform them with high efficiency and accuracy.

机器人技术

今天，即使是学校里的孩子都知道"机器人"这个词和它的含义，因为他们在展览馆、书里、卡通片和电影里见过许多机器人，真实的和虚构的。在发达国家，如日本和美国，各种工业机器人日夜不停地工作在车间里，从事着重复的、单调的、简单的和精细的工作，例如装配、涂漆、焊接、清洗、搬运、堆放、检查等。由于这些机器人是计算机化的，可重复编程，是自动的、多功能的机器，能把那些曾被认为只有人才能干的工作接管过来，并能高效地、正确地完成它们，于是这些机器人就理应得到"蓝领工人"的称号。

A robot is basically a highly automatic machine that is created to duplicate one or more physical capabilities of a person. It is a product of comprehensive techniques concerning automatic control, electronics, mechanics, computer science and other applied sciences. Like other automatic machines, robots are designed and made to help us do our work faster, better, and easier, without much intervention from human operators. The difference between those existing automatic machines and robots lies in that robots can perform their jobs more flexibly and in a more human-like manner, besides, the scope of jobs they can take is universal. Even in terms of mechanical structure, most robots are made with a number of parts mimicking human anatomy or that of some animals or insects, but they do not necessarily resemble a man.

从根本上讲，机器人是一种高自动化的机器，用来取代人类从事各种简单重复的体力劳动。它是一种综合运用了自动化控制、电子学、机械学、计算机科学和其他运用科学的技术的产品。像其他自动化机械产品一样，机器人常被设计和制造成为能帮助人们更快、更好且更容易完成工作的东西，且不受太多的人为因素干扰。与那些现有的自动化机械产品相比，机器人能够承担像人一样更灵活、更人性化的工作，而且，它的工作范围几乎是无所不包的。就其机械结构而言，大多数机器人制造都是模仿人类或一些动物和昆虫某一部分结构的，但并不需要完全像人。

There are two basic types of robots now widely used in industry and service. The first type is the fixed one with a movable arm, commonly known as the manipulator. The second type is generally referred to a mobile robot.

现在有两种基本类型的机器人在工业和服务业广泛运用：第一种类型是安装有移动手臂，通常被人们称为机械手的机器人；第二种类型是移动式机器人。

The typical industrial robots consists of 4 major components: the manipulator arm, the end effector, the actuator, and the controller. The manipulator arm is the major working part of a robot. It is reasonable to design and make a manipulator arm to rcscmble and mimic a human arm/hand. It often has a shoulder joint, an elbow and a wrist. The end effector attaches itself to the end of the robot arm. It is the device intended for performing the designed operations as a human hand can. One of the most popular types of end effector is a gripper. In its simplest form, the gripper is a pinch-type clamp that simulated a thumb and a finger. The actuator is the power source for moving the robot arm, controlling the joints and operating the end effector. Three basic types of power

典型的工业机器人包括 4 个主要部件：机械手、终端操作机构、驱动器和控制器。机械手是机器人主要的操作部分。通常设计成与人类手臂类似的外型，包括机器人肩、肘、腕 3 个部分。终端操作机构在机器人手臂末端，它是一个像人手一样承担已设计好的操作的执行部件。最流行的终端操作机构之一就是抓取器，其中最简单的抓取器的模式便是类似人手的由拇指和手指组成的捏式钳子。驱动器是机器人手的动力来源，控制连接部分和终端操作机构，其动力源主要有 3 种基本类型：电力、水力、气动力。控制器负责发送信号给动力源并

sources are used: electric, hydraulic, and pneumatic. The controller sends signals to the power source to move the robot arm to a specific position and to actuate the end effectors. Most controllers are electronic circuits in a form of NC (numerical control) or SC (servo control). They are reprogrammable to send out sequences of instructions for all movements and actions to be taken by the robot. The simplest form of the control system is an open-loop controller which controls the robot only by following the predetermined step-by-step instructions without any reference to feedback.

驱使机器人手到达准确的位置，激活终端操作机构的运行。大多数控制器是以 NC（数字控制）和 SC（电子控制）的形式出现的。它们重复编程，发送连续的指令驱动机器人的工作。最简单的控制系统是只能控制机器人按照预设的命令工作而没有任何反馈记忆的开闭控制器。

Now, intelligent robots appear more and more in different fields. In unmanned workshops, intelligent robots will be the major work force taking delicate, boring, and dangerous jobs, replacing human workers. In some environments, such as underground, underwater and outer space, intelligent robots will be the only laborers working tirelessly.

而今，智能机器人出现在越来越多的领域。在无人车间里，智能机器人将成为主要的劳动力，做精细、令人厌烦和危险的工作，从而取代工人。在某些环境中，例如地下、水下和外层空间，智能机器人将是不知疲倦的、唯一的工作者。

2.7 术语简介

BIOS：基本输入输出系统，是集成在主板上的一个 ROM 芯片，其中保存有计算机重要的输入输出程序、系统信息设置、开机自检程序和开机自举程序。

CMOS：是计算机主板上一块可读写的 RAM 芯片，用来保存当前系统的硬件配置和用户对某些参数的设定。它由主板电池供电，即使关闭机器，信息也不丢失。

GUI：图形用户界面，如 Windows 桌面，直观明了，操作方便，是应用程序追求的界面。

LAN：局域网，一种在小区内使用的网络，最远的两台计算机之间的距离一般不超过 10 公里，其数据传输速率高（10Mbit/s ~ 1000Mbit/s），误码率低，易维护，易管理，组网容易。例如，网吧、学校、机房等网络均是局域网。

WAN：广域网，也叫远程网，覆盖几十千米到几千千米，跨国家、省、市，使用电话、微波、卫星等传输信号，传输速率低，如互联网。

PCI：一种不依附于某个具体处理器的局部总线，是 CPU 和外设之间的一级总线，插槽为白色。

Exercises（练习）

1. Match the explanations in column B with words and expressions in column A.（搭配每组中同意义的词或短语）

A	B	A	B
基本输入输出系统	POST	IP 地址配置命令	Ping
CMOS 设置程序	PCI	软盘	MD/MDKIR
加电自检程序	BIOS	发送 echo 测试包命令	IPconfig
自检鸣叫错误	Flash	跟踪路由命令	floppy disk
互连外围设备	CMOS Setup	磁盘格式化命令	Telnet
闪速存储器	Copy	建立子目录命令	Tracert
文件拷贝命令	POST beep	远程登录命令	Format

2. Choose the proper words to fill in the blanks.（选词填空）

(1) _______ was the first widely-installed operating system on personal computers.

(2) To command to check a diskette for flaws and creates a directory where all the names of the diskette's files will be stored. It is _______.

(3) While there are several useful programs for analyzing connectivity, unquestionably ______ is the most commonly used program.

(4) Whenever you turn on your computer, the first thing you see is the _______ software doing its thing.

(5) The order that BIOS will try to load the operating system, it is ________.

> Words to be chosen from:（可选词）
> Ping command　　Format　　Boot Sequence　　DOS　　BIOS

3. Read the following sentences and write "T" for true and "F" for false.（对与错）

(1) Cannot find or initialize the floppy drive controller or the drive. Make sure the controller is installed correctly. (　　)

(2) Typically, Device Manager is used to check the status of computer hardware and update

device drivers on the computer. (　　)

(3) If the status is Disabled, that is not usually the result of user's action, but means that the device has a problem. (　　)

(4) If you lose the password you will be locked out of your own system, and then you will have to resort to techniques. (　　)

(5) The typical industrial robots consists of 3 major components: the manipulator arm, the end effector, and the actuator. (　　)

4. Translate the following English into Chinese and Chinese into English. （英汉互译）

Computer　Networks

Computer networks link computers by communication lines and software protocols, allowing data to be exchanged rapidly and reliably.

____

最初，网络用来提供计算机终端访问和计算机之间的文件传输的功能，如今，网络提供传递电子邮件、访问公共数据库和电子公告板的功能，并已开始用于分布式系统。

____

____

Networks also allow users in one locality to share expensive resources, such as printers and disk-systems.

____

5. Practical assignment: practice on a computer, dealing with computer system.（实践作业：上机进行计算机实践操作，熟悉计算机系统维护。）

附文 2: Reading Material（阅读材料）

How Computer Viruses Work

A computer virus—an unwanted program that has entered your system without you knowing about it—has infector and detonator. They have two very different jobs. One of the features of a computer virus that separates it from other kinds of computer program is that it replicates itself, so that it can spread (from computer to computer, or networks) to other computers.

After the infector has copied the virus elsewhere, the detonator performs the virus's main work.

Generally, that work is either damaging data on your disks, altering what you see on your computer display, or doing something else that interferes with the normal use of your computer.

Don't worry too much about viruses. You may never see one. There are just a few ways to become infected that you should be aware of. The sources seem to be service people, pirated games, putting floppies in publicly available PCs without write-protect tabs, commercial software (rarely), and software distributed over computer bulletin board systems (also quite rarely, despite media misinformation). Many viruses have spread through pirated—illegally copied or broken—games. This is easy avoided. Pay for your games, fair and square. If you use a shared PC or a PC that has public access, such as one in a college PC lab or a library, be very careful about putting floppies into that PC's drives, carry a virus-checking program and scan the PC before letting it write data onto floppies.

Despite the low incidence of actual viruses, it can't hurt to run a virus checking program now and then. There are actually two kinds of antivirus programs: viruses shields and virus scanners. Viruses are something to worry about, but not a lot. A little common sense and the occasional virus scan will keep you virus-free. Remember these four points:

Viruses can't infect a data or a text file.

Before running an antivirus program, be sure to cold-boot from a write-protected floppy.

Don't boot from floppies except reliable DOS disks or your original production disks.

Stay away from pirated software.

Future Memories

Chip makers are searching for cheap, fast, "universal" memories to replace DRAM, SRAM and flash.

For 40 years, system designers have seen memory densities double every 18 ~ 24 months while memory chip prices have remained essentially flat, cutting the cost per bit in half each time. As the technical challenges of building ever-smaller memory cells in silicon have increased, however, some memory manufacturers have predicted that the cost curve will start to swing the other way before the end of the decade.

Researchers are working on several alternate technologies that could eventually replace those in the three memory types commonly used today: Low-cost dynamic RAM (DRAM), used in PCs and servers; fast static RAM (SRAM), used for processor caches and mobile devices; and nonvolatile flash memory, used in everything from computer BIOSs to cell phones.

Researchers at IBM, Intel Corp. and other companies envision the development of a universal memory technology that could someday replace all three.

A universal memory technology could change how computers are designed. For example, nonvolatile RAM could allow computers to boot up and power down instantly because stored information wouldn't be lost when power was. But the emergence of a universal memory technology is probably at least 10 ~ 15 years away.

Others argue that a universal memory isn't possible because one memory type can't satisfy all needs. For example, nothing could be the fastest and cheapest at the same time.

Most research today is focused on addressing the limitations of one memory technology at a time, such as flash. But some attributes of today's technologies will be hard to beat.

Most technologies will probably not be able to compete on lowest cost per bit against DRAM or with the fastest SRAM. So they will fall into that space in between. Ferroelectric RAM (FRAM) and magneto resistive RAM (MRAM) are the best-funded and most-evolved of the emerging memory technologies. FRAM is a nonvolatile RAM that was developed by Ramtron International Corp. in Colorado Springs. It's licensed by Texas Instruments Inc and others. More than 30 million products have already shipped using FRAM, including metering, radio frequency identification and smart-card devices.

FRAM, which is based on nanoscale "quantum dots", uses less power and writes faster than DRAM or flash, and it has a long life span. But the technology remains 20 ~ 50 times more expensive per bit than DRAM, and chip density is far lower. Ramtron is prototyping 1Mbit parts today and hopes to push the technology to 4Mbit or 8Mbit in 2006. Until MRAM is ready for the market, however, FRAM is the only game in town for nonvolatile DRAM.

RAM

Chapter 3 Foundation of Computer Network（计算机网络基础）

教学要求

掌握专业关键词汇（key words）；能阅读本章所列英语短文；能识别计算机网络的各组件。

教学内容

计算机网络英语；各相关硬件设备品牌、主要生产厂商；常用的专业术语。

教学提示

到学校机房或有关网络公司参观，感受本章内容，以学到更多的专业知识和词汇。

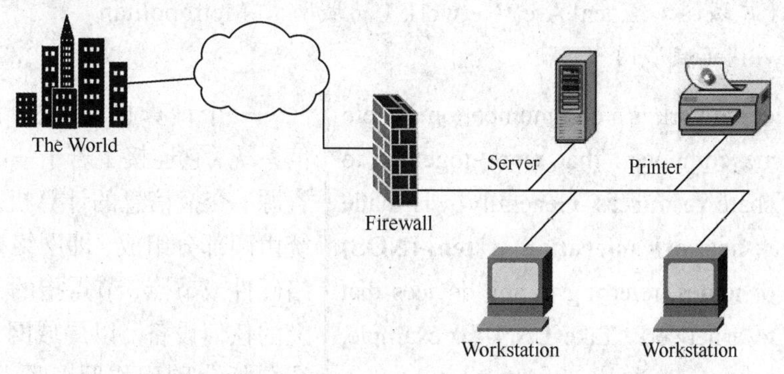

Network Design

3.1 Network Concepts（网络基础）

Networking arose from the need to share data in a timely fashion. Personal computers are wonderful business tools for producing data, **spreadsheets**, graphics, and other types of

计算机网络的建立是为了满足人们以即时方式共享数据的需求。个人计算机在处理数据、电子表格、图形以及其他类型的信息方面是理想的办公设

information, but do not allow you to quickly share the data you have produced.

备，但却不支持快速（用户输出的）数据共享。

If the user were to connect his computer to other computers, he could **share** data on other computers, including high-quality printers. A group of computers and other devices connected together are called a **network**, and the technical concept of connected computers sharing resources is called networking.

如果用户能把他的计算机与其他计算机联系在一起的话，他就可以共享其他计算机上的数据，包括高性能的打印机。一组计算机和其他的设备连接在一起构成的系统叫作网络，互连的计算机共享资源的技术（概念）叫作网络技术。

Computers that are part of a network can share data, **messages**, **graphics**, printers, **fax machines**, **modems**, **CD-ROMs**, hard disks, and other **data storage equipment**.

联网的计算机可以共享数据、消息、图形、打印机、传真机、调制解调器、只读光盘、硬盘以及其他数据存储设备。

According to the scope the network works and the distance for the mutual connection of computers, it has three categories: **Wide Area Network**, **Local Area Network,** and **Metropolitan Area Network**.

按照网络作用的范围与计算机相互连接的距离分，有广域网、局域网和城域网 3 种类型的网络。

Key words: networking（网络技术），spreadsheet（电子表格），share（共享），network（网络），message（信息），graphics（图形），fax machine（传真机），modem（调制解调器），CD-ROM（只读光盘），data storage equipment（数据存储设备），Wide Area Network（广域网），Local Area Network（局域网），Metropolitan Area Network（城域网）

A computer network is a communication system connecting some computers that work together to exchange and share **resources**. Generally it is made up of two parts: **network operation system (NOS)** and hardware or **nodes**, referring to any devices that are connected to a network. Take LAN for example, the hardware system includes a microcomputer, which is used as file server and **workstation**, net interface card, **T-shape juncture**, **BNC joint**, **terminator**, and **cables** etc.

一个计算机网络实际上是一个通信系统，它连接了若干一起工作并分享资源、交换信息的计算机。通常网络系统由两部分组成，即网络操作系统和网络硬件或节点，节点指的是连接在网络上的任何设备。以局域网为例，硬件系统包括分别用作服务器与工作站的微型计算机、网卡、T 形接头、BNC 接头、终端电阻以及电缆线等。

File Server is the soul of a whole network, so it must be the best. All the input and output of data are under control of the File Server on the network. A network must have a computer as the network File

服务器（file server）是整个网络的灵魂，所以它必须是最好的。网络上所有数据的进出都须通过服务器来控制。一个网络必须有一台计算机来做网络

Server, and it provides the workstation its data on the hard disk.

Workstation actually is a set of PC. When it has been connected with File Server and been **logged on**, it can access data from the File Server, and operates on the workstation with the documents needed.

Net card **(network interface card: NIC)** is the **interface** between File Server and Workstation.

服务器，由它将硬盘上的数据提供给工作站使用。

工作站（workstation）实际上就是一台 PC，当它与服务器连接并登录后，就可以访问服务器上的数据，在得到所需文件后就可以在工作站上运行。

网卡（network interface card: NIC）是服务器与工作站之间的接口。

 **Key words:** resource（资源），network operating system（网络操作系统），node（节点），workstation（工作站），T-shape juncture（T 形头），BNC joint（BNC 接头），terminator（终结器），cable（电缆），file server（文件服务器），log on（登录），network interface card（网卡），interface（界面，接口）

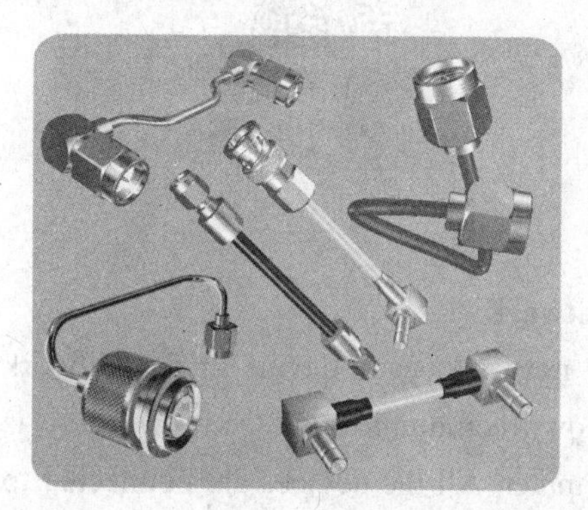

3.2 Architecture of Computer Networks（计算机网络结构）

The configuration, or the physical layout, of the **equipment** in a communication network is called **topology**. Devices connected to a network, such as **terminals**, printers, or other computers, are referred to as nodes. The four configurations or Network are star, bus, ring, and tree.

Star Network

A star network contains a central unit, a number of personal computers, terminals or **peripheral devices**. All of them are linked to the central unit either by point-to-point or multidrop cable lines. This central unit may be a **host computer** or a file server. All communications in the network are controlled through this central unit by polling. That is, before one

拓扑，指通信网中设备的配置或物理布局。节点，指连接在网络上的设备，如终端、打印机或其他计算机。通信网络的设备配置称为拓扑，主要有 4 种配置方式：星形、总线形、环形和树形。

星形网络

星形网络由一台中央处理器、多台个人计算机、终端或外围设备构成。所有这些设备都可以通过两点连接或多点连接的方式和中央处理器连接，形成星形结构。这台中央处理器可以是一台主机或是文件服务器。所有的网络通信都必须通过访问这台处理器

user wants to **send messages**, he or she must be polled or asked if there is any message to be sent. One particular advantage of star network is that it provides a **time-sharing system**. This allows several users to share resources on a central computer simultaneously.

来进行控制。也就是说，用户在传送信息之前将被询问是否有信息要传送。星形网络结构的一个显著优点是它提供分时服务，这使多个用户可以同时分享中央计算机上的资源。

Key words: equipment（设备），topology（拓扑），terminal（终端），star network（星形网络），peripheral device（外围设备），host computer（主机），send message（传送消息），time-sharing system（分时系统）

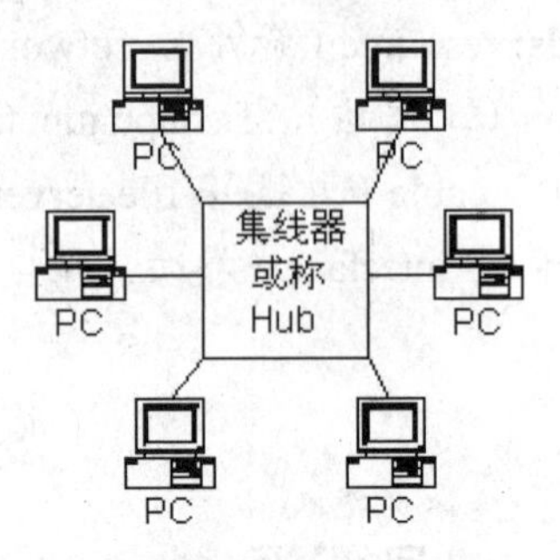

Star Network

Bus Network

In **bus network**, each device in the network handles its own communications control. There is no host computer. All the devices are connected to and share a single cable. All communication travel along this common connection cable are called **bus**. As the information passes along the bus, it's examined by each device to see if the information is intended for it.

When only a few microcomputers are to be linked together, bus network is more preferable. This configuration is common in systems for **electronic mail** or for sharing data stored on different microcomputers. The bus network is as efficient as the star network for sharing resources, because it is not directly linked to the resource.

In bus network, any devices can be attached or detached from the network at any point without disturbing the network. In addition, if one computer fails, it doesn't affect the entire network.

总线网络

在总线网络中，每一个设备处理自己的通信管理，并没有主机。所有的设备都由一根电缆线连接。所有的通信传输都沿着这根共同连接着的称为总线的电缆进行。当信息沿总线传送时，每个设备对它进行检查，看这条信息是否是对该设备的。

当想要连接少数几台计算机时，总线网络更加适合。这种配置的电子邮件或不同计算机上的数据共享系统是相同的。由于总线网络并非直接和资源连接，因此在资源共享上和星形结构一样有效。

总线网络中的设备可以连接到网络的任何一点，或从网络上任何一点取下，而不影响其他部分。而且，一台机器出故障，并不影响整个网络。

Key words: bus network（总线网络）, bus（总线）, electronic mail（电子邮件）

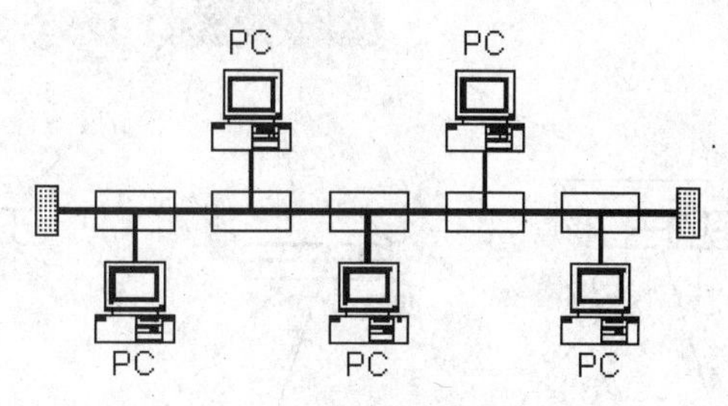

Bus Network

Ring Network and Tree Network

A **ring network** doesn't use a central host computer. Each device is connected to two other devices, forming a ring. Messages are passed around the ring until they reach the destination.

A **tree network** consists of several computers linked to a central host computer, just like a star network. However, these computers are also host to other smaller computers. It allows various computers to share **database**, processing power, and different output devices.

环形网络和树形网络

环形网络也没有主机，每个机器都和其他两台机器互相连接，形成一个环形。信息沿着环形线传送直到到达目的地。

如同星形网络一样，树形网络由若干计算机连接到一台中央计算机上。然而，这些计算机同时又成为其他更小计算机的主机。树形网络允许各种计算机分享数据库、字处理程序和其他不同的输出设备。

Key words: ring network（环形网络）, tree network（树形网络）, database（数据库）

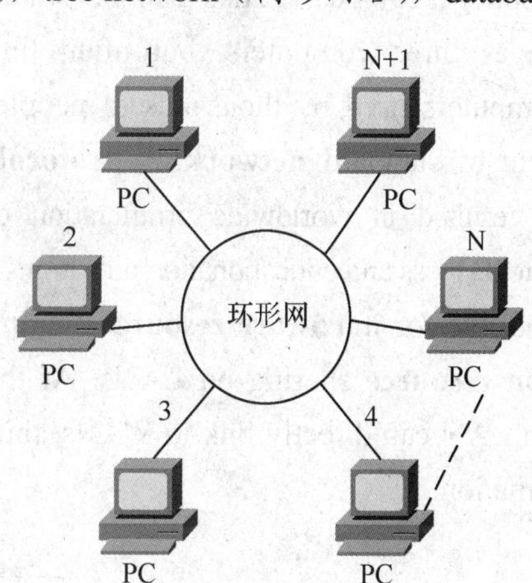

Ring Network

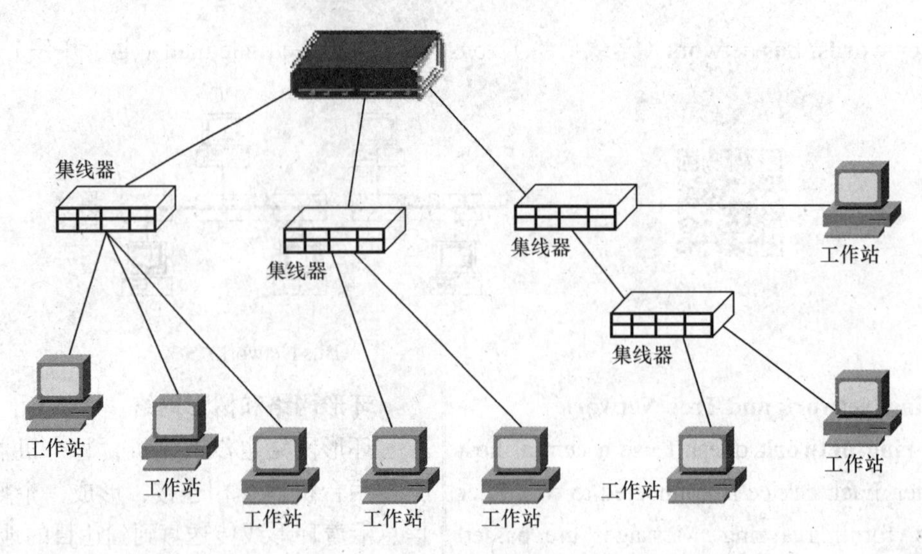

Tree Network

3.3 LAN（局域网）

A network is any collection of independent computers that **communicate** with one another over a shared **network medium**. LANs are networks usually confined to a geographic area, such as a single building or a college campus. LANs can be small, linking as few as three computers, but often link hundreds of computers used by thousands of people. The development of standard **networking protocols** and media has resulted in worldwide proliferation of LANs throughout business and educational organizations.

LAN is possible for **hardware resource** sharing and **information resource** sharing on LANs. At the same time, each LAN can directly link to WANs, thus gains rich information.

网络就是连接在一起的独立计算机，相互之间通过一个共享网络媒介进行通信的。局域网通常是指某一地域范围内的一个网络，如一幢建筑物内或一个大学校园内。局域网可以很小，可以少到只连接3台计算机，但通常会连接几百台计算机，供几千人使用。标准网络协议和网络媒介的发展使得局域网在全世界快速发展，它遍及商业领域和各教育机构。

局域网可以实现硬件资源共享和信息资源共享。此外，局域网还能直接连接到广域网上，从而获得上面的丰富信息。

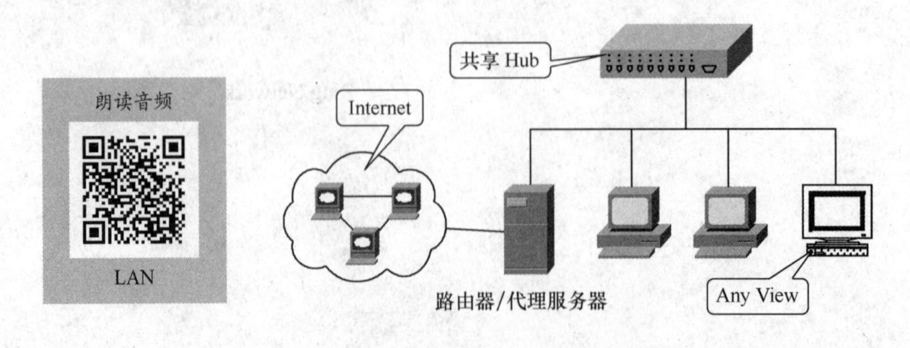

 **Key words:** communicate（通信）, network medium（网络媒介）, network protocol（网络协议）, hardware resource（硬件资源）, information resource（信息资源）

What Is Wireless Computer Networking?

Wireless networks utilize **radio waves** and/or **microwaves** to maintain **communication channels** between computers. Wireless networking is a more modern alternative to **wired networking** that relies on copper and/or fiber optic cabling between **network devices**.

A wireless network offers advantages and disadvantages compared to a wired network. Advantages of wireless include mobility and elimination of unsightly cables. Disadvantages of wireless include the potential for radio interference due to weather, other **wireless devices**, or obstructions like walls.

Wireless is rapidly gaining in popularity for both home and business networking. Wireless technology continues to improve, and the cost of **wireless products** continues to decrease. Popular **wireless local area networking (WLAN)** products conform to the 802.11 "Wi-Fi" standards.

The gear a person needs to build wireless networks includes network adapters (NICs), access points (APs), and routers.

什么是无线计算机网络？

无线网络利用无线电波和（或）微波保持计算机之间的通信通道。无线网络比依赖于铜线和（或）光缆来连接网络设备的有线网络更现代些。

无线网络与有线网络相比既有优点也有缺点。其优点是移动性强，并且除去了难看的电缆线。其缺点是环境带来的潜在的无线电干扰，如有天气原因，或其他无线上网设备或障碍物（如墙）的阻挡都可能带来干扰。

无线网络在家庭网络和商业网络两方面迅速流行起来。无线技术还在继续提高和完善，而无线产品的价格却在不断降低。一般的无线局域网（WLAN）产品都遵循 802.11 "Wi-Fi" 标准。

一个人要建立无线网络必须有网络适配器（NICs），访问端口和访问通道。

 **Key words:** wireless computer networking（无线计算机网络）, radio wave（无线电波）, microwave（微波）, communication channel（通信通道）, wired networking（有线网络）, network device（网络设备）, wireless device（无线设备）, wireless product（无线产品）, wireless local area networking（WLAN，无线局域网）

 无线网关键字

802.11x：定义了无线局域网技术的一系列标准。此标准由电气电子工程协会（IEEE）制定，它包括 802.11a、802.11b、802.11g 和新兴 802.11n。802.11ac 标准在本文发布时仍在开发中。

Wi-Fi：代表"无线保真"，指 802.11 标准的 IEEE802.11b 子集。Wi-Fi 支持高达 11Mbit/s 的数据传输率，是迄今为止最常用的标准。

WLAN：无线局域网的缩写，指采用 802.11 无线技术进行互连的一组计算机和相关设备。也称为 LAWN。

WAN：广域网，指覆盖主要城市中心或整个国家的网络，如 GPRS、CDMA 和 GSM 等。

无线热点：通过无线接入点为移动用户提供连网或互联网服务的区域。公共热点建立

在图书馆、机场、酒店甚至咖啡厅和餐馆中。

蓝牙：一种低成本、低功率无线“线缆替代”技术，使电话、笔记本电脑、PDA 和外设等设备能够通过短程（10 米）无线信号进行互连。

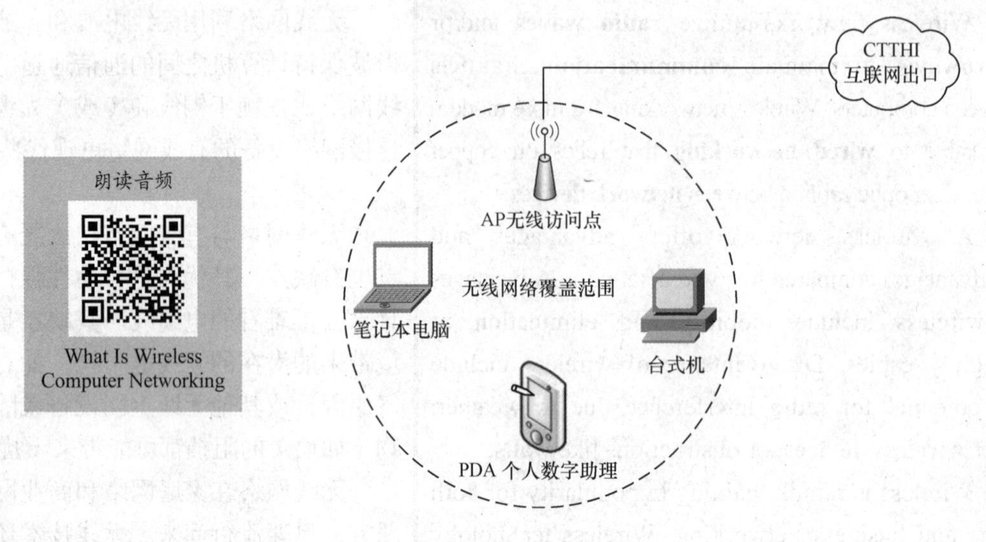

3.4 WAN（广域网）

As the geographical scope of the network grows by connecting users in different cities or different states, the LAN grows into a wide area network (WAN).

WAN (wide area network) is also called **RCN (remote computer network)**. Its effective scope can reach from tens of kilometers to ten thousands of kilometers. A wide area network uses **telephone lines**, microwaves, satellites, or a combination of **communication channels**. The network between nations belongs to wide area network. At present the biggest wide area network in the world is Internet and it has covered more than 220 nations. It is like a giant highway that connects you to millions of other people and organizations located throughout the world. The **Internet** is a huge computer network available to nearly everyone with a microcomputer and a means to connect to it. The **Web**, also known as the **World Wide Web** or **WWW**, is an **Internet service** that provides a multimedia interface to numerous resources available on the Internet.

随着各个城市、各个国家的用户被连接在一起，网络的地理覆盖范围不断扩展，局域网（LAN）已经发展成了广域网（WAN）。

广域网也称远程网 RCN（remote computer network），其作用范围可以达几十公里到几万公里。广域网使用电话线、微波、卫星或这些通信信道的组合来传送信息。国家与国家之间的网络就属于广域网。目前世界上最大的广域网就是互联网，其范围已经覆盖了全球 220 多个国家。它像一条高速公路一样将你与遍布全世界的多达几百万的人和机构连接起来。互联网是一个巨大的计算机网络，它适合几乎每一个有计算机和上网条件的人去使用它。环球网，也就是众所周知的万维网，是互联网的一项服务，它为互联网上大量的可用资源提供一个多媒体的界面。

 **Key words:** RCN（remote computer network，远程网），telephone line（电话线），communication channel（通信信道），Internet（互联网），Web（环球网），World Wide Web（万维网），Internet service（互联网服务）

Internet

The Internet is a system of linked networks that are worldwide in scope and facilitate data communication services such as **remote login**, **file transfer**, electronic mail, the World Wide Web, and **newsgroups**.

With the meteoric rise in demand for connectivity, the Internet has become a communication highway for millions of users. The Internet was initially restricted to military and academic institutions, but now it is a full-fledged conduit for any and all forms of information and commerce. Internet **websites** now provides personal, educational, political, and economic resources to every corner of the planet.

互联网

互联网是一个全球网络系统，它有许多方便的数据通信服务，如远程登录、文件传输、电子邮件、万维网和新闻组等。

随着联网需求的迅速上升，互联网已成为几百万用户间的信息通信高速公路。互联网最初仅限于军事部门和学术机构使用，而现在它已经是一个完全成熟的通信通道以及信息和商业的窗口。现在互联网各网络站点向全球每一个角落提供个人、教育、政治和经济等资源。

 **Key words:** remote login（远程登录），file transfer（文件传输），newsgroups（新闻组），website（网络站点）

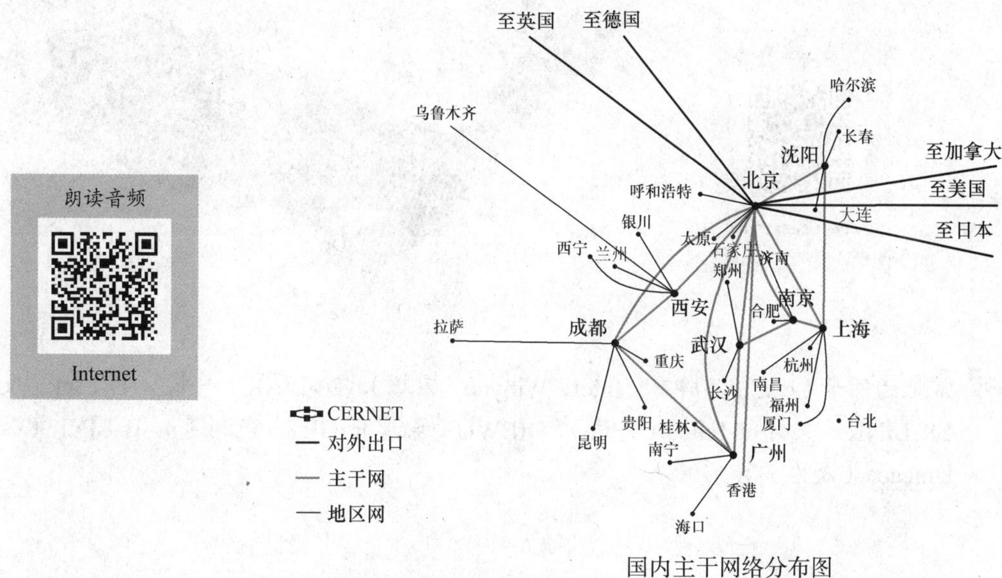

国内主干网络分布图

中国的计算机网络主要有六大互联网络：中国科学技术网（CSTNet）、邮电部中国公用计算机互联网（ChinaNet）、中国教育和科研计算机网（CERNet）、中国金桥信息网（ChinaGBN）、中国联通工联网（UNINET）和中国网通（CNCNET），并实现了网络间的互联互通。中国互联网就是以这六大互联网为依托而产生和发展起来的。

3.5 Hardware Concepts of Computer Network（计算机网络硬件设备基础）

3.5.1 Network Interface Card（网卡）

A network interface card **(NIC)** is a computer circuit board or card that is installed in a computer so that it can be connected to a network. Personal computers and workstations on a local area network (LAN) typically contain a network interface card specifically designed for the LAN **transmission technology**, such as **Ethernet** or **Token Ring**. Network interface cards provide a dedicated, full-time connection to a network. Most home and **portable computers** are connected to the Internet through as-needed **dial-up connection**. The modem provides the connection interface to the **Internet service provider**.

网卡是安装在计算机内的一块电路板（或插卡），计算机通过它连接到网络上。在局域网中的个人计算机和工作站中都会有一块专为局域网传输技术而设计的网卡，如以太网或令牌环网。网卡专用于全时连接网络。绝大多数家用电脑和手提电脑通过拨号连接上网。调制解调器为互联网服务提供商提供了一个连接接口。

Key words: NIC（网卡）, transmission technology（传输技术）, Ethernet（以太网）, Token Ring（令牌环网）, portable computer（手提电脑）, dial-up connection（拨号连接）, Internet service provider（互联网服务提供商）

常见的网卡（NIC）品牌有：Intel、Winyao（万耀）、TP-LINK（普联）、D-Link（友讯）、LR-LINK（联瑞）、Tenda（腾达）、DIEWU（蝶舞）、GRT（光润通）、B-LINK（必联）、Unicaca（众力）等。

3.5.2 Modem（调制解调器）

Modem is short for "**modulator-demodulator**". A modem is a device or program that enables a computer to transmit data over, for example, telephone or cable lines. Computer information is stored **digitally**, whereas information transmitted over telephone lines is transmitted in the form of analog waves. A modem converts between these two forms.

Fortunately, there is one **standard interface** for connecting **external modems** to computers called RS-232. Consequently, any external modem can be attached to any computer that has an RS-232 port, which almost all personal computers have. There are also modems that come as an **expansion board** that you can insert into a vacant **expansion slot**. These are sometimes called onboard or **internal modems**.

调制解调器（Modem）是"调制—解调（modulator，demodulator）"的缩写。调制解调器是一个帮助计算机转换（比如电话线、电缆线）数据的设备或程序。计算机中的信息是以数字方式存储的，而在电话线中传输的却是模拟信号。调制解调器就将这两种信号互相转换。

幸好，外置调制解调器都有一个叫作 RS-232 的标准接口与计算机相联，因此，任何一个外置调制解调器都能连接到任何一台有 RS-232 端口的计算机上，而几乎所有的个人计算机都有这种端口。也有的调制解调器以扩展板的形式插入空的扩展槽当中。有时它们被称为内置调制解调器。

Key words: modulator-demodulator（调制解调器），digitally（数字地），standard interface（标准接口），external modem（外置调制解调器），expansion board（扩展板），expansion slot（扩展槽），internal modem（内置调制解调器）

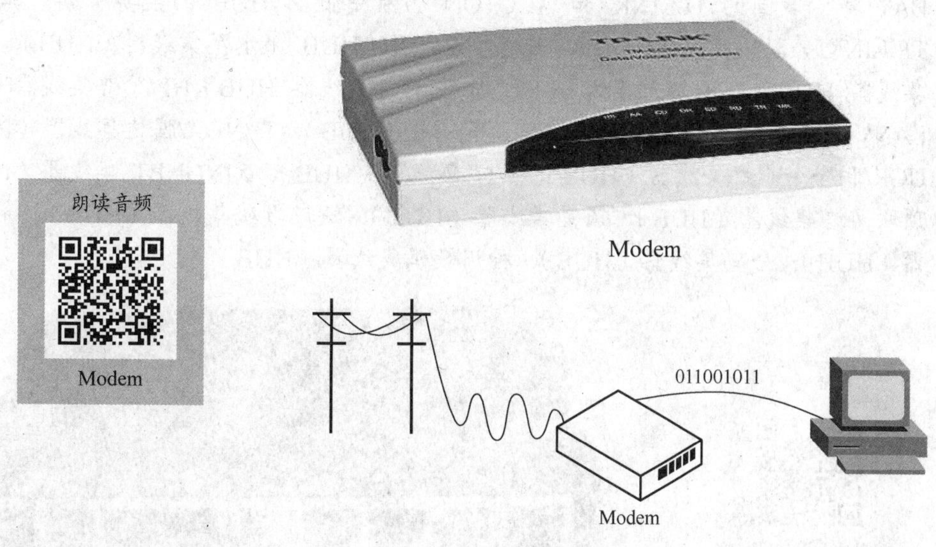

3.5.3 Hubs（集线器）

A **hub** is a common **connection point** for devices in a network, which is commonly used to connect

集线器是网络设备的一个普通连接点，它通常用于连接局域网中的各网

segments of a LAN. A hub contains multiple ports. When a **packet** arrives at one **port**, it is copied to the other ports so that all segments of the LAN can see all packets.

A **passive hub** serves simply as a conduit for the data, enabling it to go from one device (or segment) to another. So-called **intelligent hubs** include additional features that allow an administrator to monitor the **traffic** passing through the hub and to **configure** each port in the hub. Intelligent hubs are also called **manageable hubs**.

A third type of hub, called a **switching hub**, actually reads the **destination address** of each packet and then forwards the packet to the correct port.

段。一个集线器有若干个端口。当一个数据包到达一个端口时，它将被复制到其他端口，因此局域网中所有网段都能看到所有的数据包。

一个被动集线器仅仅是一个数据通道，它使数据从一个设备（或网段）传输到另一个上面。所谓的智能集线器，还包括其他的特征，即允许管理员监控通过集线器的流量和配置集线器的每个端口。智能集线器也被称为可控集线器。

第三种集线器被称为交换集线器，事实上是先读取每个数据包中的目的地址，然后将数据包传送到正确的端口。

Key words: hub（集线器），connection point（连接点），segment of a LAN（局域网段），packet（信息包，数据包），port（端口），passive hub（被动集线器），intelligent hub（智能集线器），traffic（通信量，流量），configure（配置），manageable hub（可控集线器），switching hub（交换集线器），destination address（目的地址）

常见的集线器（HUB）品牌：高档 HUB 主要由美国品牌占领，如 3COM、INTEL、BAY 等；我国的 D-LINK 和 ACCTON 占有中低端 HUB 的主要份额；其他的有 TP-LINK/普联集线器（HUB）、优越者集线器（HUB）、水星集线器（HUB）、阿尔法集线器（HUB）、SSK/飚王集线器（HUB）、ECOM 集线器（HUB）、HP/惠普集线器（HUB）、LANVOLAN 集线器（HUB）、TCL 集线器（HUB）、TENDA/腾达集线器（HUB）、UGR/联合金彩虹集线器（HUB）、WG 集线器（HUB）、XINGNET 集线器（HUB）、顶联网络集线器（HUB）、顶星集线器（HUB）、捷讯数码集线器（HUB）、奇胜集线器（HUB）、全码集线器（HUB）、神州数码集线器（HUB）等。

Hub

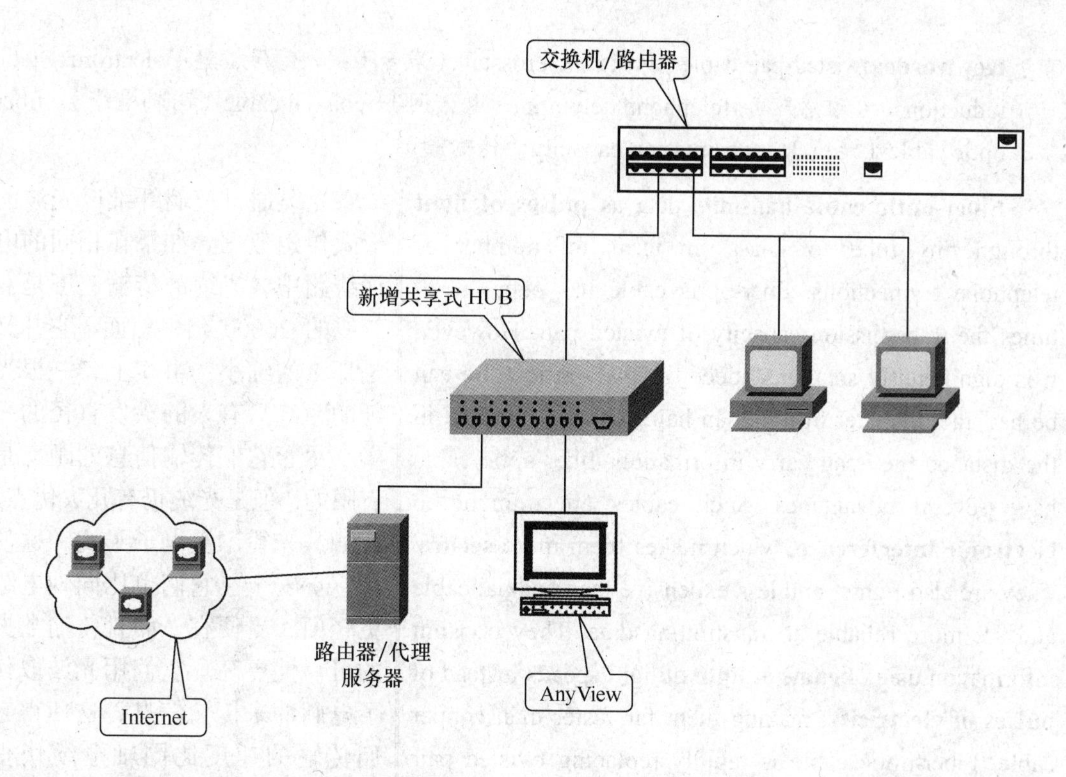

3.5.4 Network Medium（网络介质）

Twisted-pair cable, a type of cable that consists of two independently insulated wires twisted around one another. The use of two wires twisted together helps to reduce **crosstalk** and **electromagnetic induction**. While twisted-pair cable is used by older **telephone networks** and is the least expensive type of local-area network (LAN) cable. Most networks contain some twisted-pair cabling at some point along the network. Other types of cables used for LANs include **coaxial cables** and **fiber optic cables**.

双绞线，一种由两根独立的绝缘金属丝相互缠绕在一起构成的电缆。两根金属丝相互缠绕在一起是为了降低相互间的信号串扰和电磁感应。虽然双绞线常用于老的电话网和费用低廉的局域网中，但在绝大多数的网络分支上也用到了双绞线。局域网中用到的其他类型的电缆还有同轴电缆和光缆。

Coaxial cable, a high-frequency transmission cable, replaces the multiple wires of telephone lines with a single solid-copper core. In terms of number of telephone connections, a coaxial cable has over 80 times the **transmission capacity** of twisted pair. Coaxial cable is used to deliver television signals as well as to connect computers in a network.

同轴电缆，一种高频传输电缆，用一根单一的铜芯取代了多金属丝的电话线。如果在相同的电话线路中，同轴电缆的传输速度是双绞线的 80 倍。同轴电缆也常被用于传送电视信号，以及在网络中连接计算机。

Key words: twisted-pair cable（双绞线），crosstalk（串音干扰，交调失真），electromagnetic induction（电磁感应），telephone networks（电话网），coaxial cable（同轴电缆），fiber optic cable（光缆），transmission capacity（传输量）

Fiber-optic cable transmits data as **pulses of light** through tiny tubes of glass. In terms of number of telephone connections, fiber-optic cable has over 26000 times the transmission capacity of twisted pair. However, it is significantly smaller. Indeed, a **fiber-optic tube** can be half the diameter of a human hair. Although limited in the distance they can carry information, fiber-optic cable have several advantages. Such cables are immune to **electronic interference**, which makes them more **secure**. They are also lighter and less expensive than coaxial cable and are more reliable at transmitting data. They transmit information using **beams of light** at light speeds instead of **pulses of electricity**, making them far faster than copper cable. Fiber-optic cable is rapidly replacing twisted pair telephone lines.

光缆通过玻璃纤维以光波的形式传送数据。如果在相同的电话线路中，光缆的传输速度是双绞线的 26000 倍。然而，它却又是那么的细小，事实上，一根光纤的粗细只有人的头发直径的一半。尽管它们传输信息的距离是有限的，但是光缆仍有几大优点。比如，光缆不受电子干扰，这使得它更安全。它们也比同轴电缆更轻便、更便宜，而且传输数据的可靠性更高。它们用光波取代了电脉冲来传输信息，这使得它们传输的速度比同轴电缆快得多。在电话线路中，光缆迅速地取代了双绞线。

Key words: pulse of light（光脉冲），fiber-optic tube（光纤），electronic interference（电子干扰），secure（安全），beam of light（光波），pulse of electricity（电脉冲）

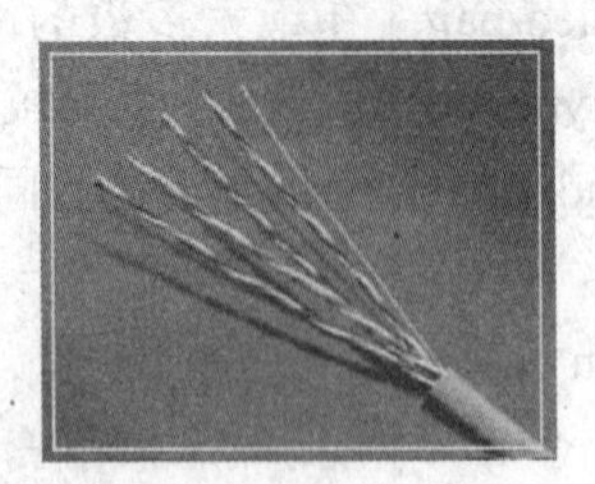

Twisted-pair Cable

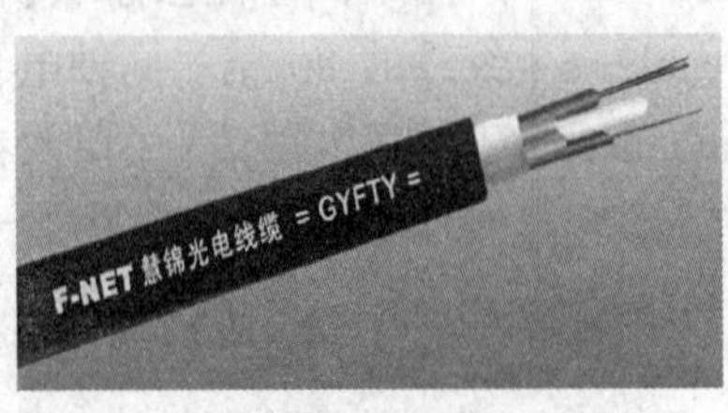

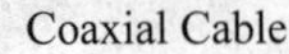

Coaxial Cable

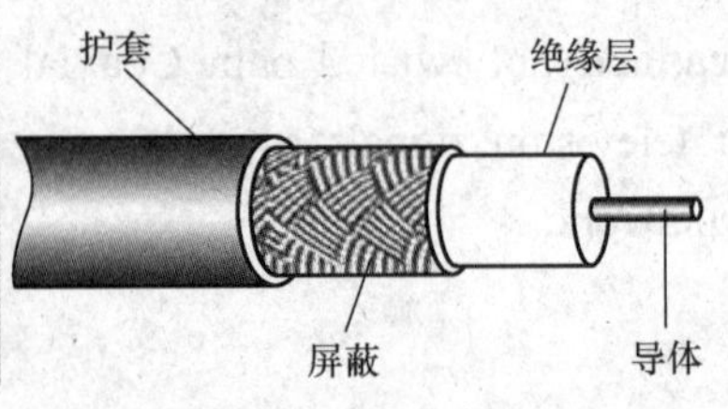

Fiber-optic Cable

3.6 Network Test Tools（网络测试命令）

Problems happen! Even when the network is monitored, the equipment is reliable, and the users are careful, things may go wrong. The test of a good network administrator is the ability to analyze, troubleshoot, and correct problems under pressure of a network failure that causes company downtime. Besides the processes that are described by some veterans, there are software tools that are available for you to use to solve network connectivity problems. These tools can help in LAN troubleshooting, but are especially helpful in a WAN troubleshooting situation. These commands include the Ping, Tracert, Telnet, Netstat, ARP, and IPconfig, which are run in command-line mode.

问题来了！就算我们监视着整个网络，设备可靠，并且用户很小心，计算机也可能会发生故障。迫于网络故障导致的公司停工的压力，一个优秀的网络管理员应具备分析、检测和解决网络故障的能力。除了某些经验丰富的老手所描述的处理步骤外，还可借助有效的软件工具来解决网络连接问题。这些工具可用于局域网故障检测，但在广域网中更加有用。它们包括 Ping、Tracert、Telnet、Netstat、Arp 和 IPconfig，都运行在命令行模式下。

Network Test Tools

3.6.1 Ping 命令

While there are several useful programs for analyzing connectivity, unquestionably Ping is the most commonly used one. The Ping program sends an Internet Control Message Protocol (ICMP is the Internet protocol that reports errors and provides information relevant to IP packet addressing) echo request packet to the destination host. When the echo request packet is received, the remote host immediately returns an echo reply packet to the sending device. The format of the Ping command is:

在几个检测网络流通性的有用程序中，毫无疑问 Ping 命令是最常用的。Ping 命令向目的主机发送互联网控制消息协议（ICMP 用来报告 IP 数据包寻址的错误和相关信息）的 echo 数据包。当 echo 包被目的端接收时，该主机立刻向源端回传应答数据包。Ping 的格式如下：

Ping 1

Ping [-t] [-a] [-n count] [-l length] [-f] [-i ttl] [-r count] destination

-t	Ping the specified host until stopped. To see statistics and continue—type Ctrl+Break; To stop — type Ctrl+C（Ping 指定的主机直到停止操作。看统计信息并继续执行该命令——按下 Ctrl+Break，停止则按下 Ctrl+C）
-a	Resolve host name and ping address（主机名解析为 IP 地址，并且 Ping 该地址）
-n	number of echo requests to send（发送 count 指明的 echo 数据包个数）
-l	length—send specified size echo packets［定义 echo 数据包的大小（byte 为单位）］
-f	DO NOT FRAGMENT command sent to gateways（在数据包中发送“禁止分段”标志，将其送至网关）
-i	ttl sets the TTL field［指定 TTL 值在对方系统停留的时间（1～255）］
-r	count records the route of the outgoing and returning packets（在“记录路由”字段中记录传出和返回数据包的路由）
destination	Specify the remote host to ping, by domain name or by IP address（指定所要 Ping 的远端主机，它可以是域名也可以是 IP 地址）

In the examples below (Experiment 1 and 2), an ECHO_REPLY was received.

下面这个实例中（实验 1 和实验 2），收到了 echo 应答包。

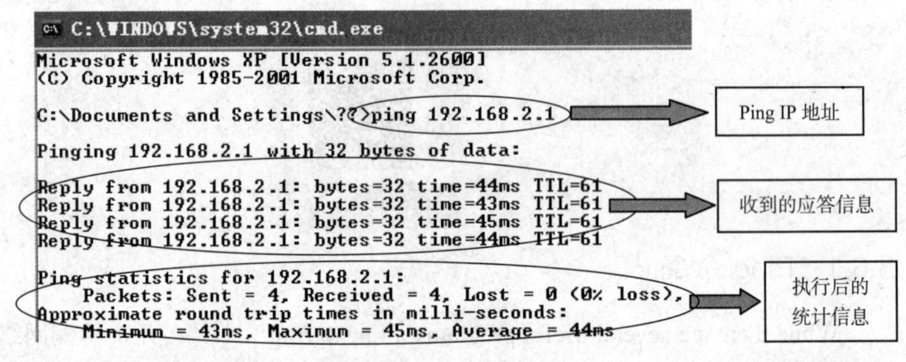

实验 1

```
C:\Documents and Settings\??>ping www.google.com

Pinging www.l.google.com [64.233.189.104] with 32 bytes of data:

Reply from 64.233.189.104: bytes=32 time=99ms TTL=244
Reply from 64.233.189.104: bytes=32 time=99ms TTL=244
Reply from 64.233.189.104: bytes=32 time=99ms TTL=244
Reply from 64.233.189.104: bytes=32 time=101ms TTL=244

Ping statistics for 64.233.189.104:
    Packets: Sent = 4, Received = 4, Lost = 0 (0% loss),
Approximate round trip times in milli-seconds:
    Minimum = 99ms, Maximum = 101ms, Average = 99ms
```

Ping 域名

实验 2

Users can also Ping the local host, this will allow you to see if the computer is able to send information out and receive the information back (Experiment 3). Make sure that this does not send information over a network

用户也可 Ping 本地主机，这样你就可以知道计算机是否能发送和接收信息，如实验 3 所示。注意该操作并非通过网络发送信息而是使

but may allow you to see if the card is being seen. Maybe you would like to Ping 127.0.0.0, the loopback address for the computer. The Ping is successful, so that eliminates a possible problem between the computer, driver configuration, and the NIC card. If the Ping operation is not able to complete, this should relay back an unsuccessful message which could be an indication of cable issues, network card issues, hub issue, etc.

你能得知网卡是否完好。也许你会Ping 127.0.0.0，这是计算机的回环地址（网卡地址）。Ping操作成功，排除了可能存在于计算机、驱动程序设置和网卡的问题。如果Ping操作未能完成，将有一条失败信息传回来，表明可能是电缆问题、网卡问题或集线器问题等。

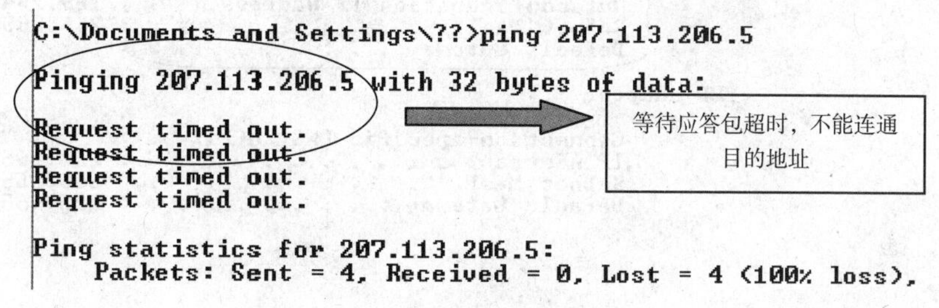

实验 3

3.6.2 IPCONFIG 命令

If you are looking for basic information quickly, Microsoft provides one of two programs for this purpose, depending on which version of Windows you use. The utility Winipcfg is included with Windows 95/98. A command-line program, IPconfig, is included with Windows NT and Windows 2000 and in Microsoft's TCP/IP stack for Windows for Workgroups. Both programs provide the same information. These Windows utilities display IP-addressing information for the local network adapter(s) or a specified NIC.

微软提供了两个程序，以便快速查找网络配置基本信息，这两个程序用哪一个取决于用户使用的Windows版本。Winipcfg是在Windows 95/98下使用的；IPconfig是一个命令行程序，包含在Windows NT、Windows 2000和微软的Windows工作组的TCP/IP堆栈里，它们都有相同的功能。这些Windows实用程序显示了本地网络适配器或专用网卡上的IP寻址信息。

Command format:

命令格式：

IPconfig [-all | -renew [adapter] | -release [adapter]]

-all	all information about adapter(s) （网络适配器的所有信息）
-renew	renew DHCP lease information for all local adapters if none is named （若所有本地网络适配器都未命名则更新DHCP配置参数）
-release	release DHCP lease information disabling TCP/IP on this adapter （禁用本地系统上的TCP/IP并发布当前的DHCP配置参数）

The following is an example of the output you should expect to see when running IPconfig (Experiment 4). For more details you can execute IPconfig -all.

下面是一个输入 Ipconfig 命令后所能看到的结果的实例，如实验 4 所示。执行 Ipconfig-all 可以得到更多的具体信息。

```
C:\Documents and Settings\??>ipconfig

Windows IP Configuration


Ethernet adapter ????:

        Connection-specific DNS Suffix  . :
        Autoconfiguration IP Address. . . : 169.254.132.121
        Subnet Mask . . . . . . . . . . . : 255.255.0.0
        Default Gateway . . . . . . . . . :

PPP adapter ??:

        Connection-specific DNS Suffix  . :
        IP Address. . . . . . . . . . . . : 220.165.186.246
        Subnet Mask . . . . . . . . . . . : 255.255.255.255
        Default Gateway . . . . . . . . . : 220.165.186.246
```

依次为：DNS 字尾细节信息
自动 IP 地址配置
子网掩码
默认网关

实验 4

These are the tools that will allow a network administrator or ordinary users to remotely monitor and control the network. As the old saying goes everything has two sides, you must utilize them on the legality ways, not the hacker's guilty way!

以上这些都是为网络管理员或普通用户提供的对网络的远程监视和控制工具。正如古谚所说，"事事皆两面"，你必须把这些工具用于合法途径，而非计算机黑客的罪恶行径！

Key words: Ping（发送 echo 测试包命令），Ipconfig（IP 地址配置命令），Time-To-Live (TTL)（生存时间）

3.7 Situation Dialogue（情景对话）

Wonderful Network

Yaoyao and Yangyang are good friends. One day, when Yangyang comes to Yaoyao's home, she is quite surprised at her chatting on computer.

Yangyang: Are you chatting on QQ?

Yaoyao: yeah, I'm going to install a web-cam, then I can chat on video.

Yangyang: The net is so nice! We've just bought a computer, but how can I log in

神奇的网络世界

姚姚和洋洋是好朋友。一天，洋洋到姚姚家玩，洋洋看见姚姚正用计算机聊天，洋洋觉得十分惊奇。

洋洋：你在用 QQ 聊天吗？

姚姚：是呀。我准备再安装一个摄像头，那样就可以视频聊天了。

洋洋：网络真好！我家也刚买了计算机，可怎样才能上网呢？

Internet ?

Yaoyao: Do you have NIC and Modem?

Yangyang: Yes, I have.

Yaoyao: Then you can install the NIC and drivers program, and apply for an ISP account.

Yangyang: So we can chat on video?

Yaoyao: Yes, we can also send E-mails and cards to each other conveniently. You only need to spend a few minutes on www.126.com or other web sites to apply for an E-mail adress.

Yangyang: How can I apply?

Yaoyao: It's quite easy. You may enter a certain web-site (such as www.163.com) and fill in the related information according to the instruction.

Yangyang: Good, I'll send you the electronic cards in holidays. Is there anything else funny online?

Yaoyao: Too much, you may read news, see films, search for materials, listen to music, read books, play games, go shopping, chat, and transmit data.

Yangyang: Can I download the learning material and find my favorite stars ?

Yaoyao: Sure, playing games on computer is also quite interesting, but you should plan well, don't spend too much time on it, otherwise it will interfere with your study.

Yangyang: I see. I like English very much. Are there many English materials on Internet?

Yaoyao: Too much! If you want to learn perfect English, you may directly enter the foreign website. There are all about English. It's quite helpful to you.

Yangyang: So wonderful!

Yaoyao: Of course. That's so-called "The

姚姚：有没有网卡和调制解调器/猫？

洋洋：都一起买了。

姚姚：那你安装好网卡和调制解调器的驱动程序后，再到电信局申请一个 ISP 账号就可以了。

洋洋：这样，我们就可以视频聊天了吗？

姚姚：我们还能互相方便地发送电子邮件和贺卡。你只须到 www.126.com 或其他的网站花几分钟时间申请一个电子邮箱（E-mail）就可以了。

洋洋：怎么申请呢？

姚姚：很简单，你进入某网站（比如 www.163.com）后，只需根据提示填写相应信息就可以了。

洋洋：好，以后的节日我会给你发电子贺卡的。在网上还有其他好玩的吗？

姚姚：太多了！如看新闻、看电影、查资料、听歌、看书、玩游戏、购物、聊天，还有数据传输……

洋洋：我可以在网上下载学习资料，并且找到我喜欢的明星吗？

姚姚：当然啦！玩网络游戏也特过瘾。不过就是要计划好时间，不能玩得太久了，否则会影响学习。

洋洋：知道了。我很喜欢英语，网上英语资料多吗？

姚姚：很多，如果你要看地道的英语，就直接进入国外的网站，里面全是英语内容，多看看对英语学习会有很大的帮助。

洋洋：太棒了！

姚姚：当然啦，所谓“世界小网络，网

world is a small network, and the network is a big world."

络大世界"嘛。

3.8 Reading and Compacting（对照阅读）

Are Wireless Networks Secure?

No computer network is truly secure, but how does wireless network security stack up to that of traditional wired networks?

Unfortunately, no computer network is truly secure. It's always theoretically possible for eavesdroppers to view or "snoop" the traffic on any network, and it's often possible to add or "inject" unwelcome traffic as well. However, some networks are built and managed much more securely than others. For both wired and wireless networks alike, the real question to answer becomes—is it secure enough?

Wireless networks add an extra level of security complexity compared to wired networks. Whereas wired networks send electrical signals or pulses of light through cable, wireless radio signals propagate through the air and are naturally easier to intercept. Signals from most wireless LANs (WLANs) pass through exterior walls and into nearby streets or parking lots.

Network engineers and other technology experts have closely scrutinized wireless network security because of the open-air nature of wireless communications.

The practice of ward riving, for example, exposed the vulnerabilities of home WLANs and accelerated the pace of security technology advances in home

无线网络安全吗？

没有哪个计算机网络是真正安全的，那么无线网络与传统有线网络相比又如何呢？

遗憾的是，没有一个计算机网络是真正安全的。从理论上讲，任何网络上的信息总是可能被偷看或探听的，并且网络中经常可能会被添加或注入一些不受欢迎的信息。然而，一些更安全的网络已经建立并运行起来了。有线网络和无线网络都一样，它们真正要回答的问题是——它是足够安全的吗？

与有线网络相比，无线网络增加了一个额外的复杂的安全级别。主要是鉴于有线网络是通过电缆来传送电信号或光波的，而无线电信号却是在空气中传播的，自然容易被截取。大多数无线网络发出的信号都是通过外墙进入附近的街道或停车场的。

由于无线通信的开放环境，网络工程师和其他技术专家已经对无线网络的安全性进行了细致的考察。

对无线电感应（比如家用无线局域网所暴露出的弱点）防护措施的推进，加速了家用无线局域网设备安全

wireless equipment.

Overall, conventional wisdom holds that wireless networks are now "secure enough" to use in the vast majority of homes, and many businesses. Security features like 128-bit WEP and WPA can scramble or "encrypt" network traffic so that its contents can not easily be deciphered by snoopers. Likewise, wireless routers and access points (APs) incorporate access control features such as MAC address filtering that deny network requests from unwanted clients.

Obviously every home or business must determine for themselves the level of risk they are comfortable in taking when implementing a wireless network. The better a wireless network is administered, the more secure it becomes. However, the only truly secure network is the one never built!

技术的步伐。

总的来看，大家普遍认为，对于广大家庭和众多的商家而言，无线网络现在已经足够安全了。像128位的WEP协议和WPA协议加密网络信息的安全特性就使得窥探者不容易解密。同样，无线网络路由器和访问点混合存取控制的特性（如物理地址过滤）也能拒绝非法用户的网络请求。

显然，当使用无线网络时，每个家庭用户和商家都必须确定他们自己的安全级别，以获得最合适的网络信息。当然一个更有效的无线网络管理，也会使它变得更安全些。然而，一个绝对安全的网络还从未建立起来。

Notes

1. WEP: short for Wired Equivalent Privacy, a security protocol for wireless local area networks (WLANs) defined in the 802.11b standard. WEP is designed to provide the same level of security as that of a wired LAN.（WEP：有线等效保密的缩写，在802.11b标准上定义的无线局域网安全协议。WEP被设计为提供与有线局域网同一水平的安全性。）

2. WPA: short for Wi-Fi Protected Access, a Wi-Fi standard that was designed to improve upon the security features of WEP.（WPA：无线保真存取保护的缩写，"无线保真"标准就是被设计来改善WEP的安全性的。）

3.9 术语简介

1. TCP/IP——传输控制协议/网际协议，通常两个协议同时用，称为TCP/IP。

2. IP address——IP地址。连接到互联网上的每台计算机必须有的唯一的地址。它由四段数字构成，又称为“数字地址”，例如168.160.233.10。

3．IPv6：其全称是“互联网协议第6版”，目前的IPv4的地址是32位编码。IPv6的地址是128位编码，能产生2的128次方个IP地址，其资源几乎是无穷的。IPv6的技术特点：①地址空间巨大：IPv6地址空间由IPv4的32位扩大到128位，2的128次方形成了一个巨大的地址空间。采用IPV6地址后，未来的移动电话、冰箱等信息家电都可以拥有自己的IP地址。②地址层次丰富，分配合理。③实现IP层网络安全。④无状态自动配置。

4．DNS——域名服务器。它负责将域名翻译成IP地址。

5．Telnet——远程登录。互联网的功能之一。它允许用户将自己的计算机远程连接到另一台计算机上，并运行计算机上的各种程序。

6．E-mail——电子邮件。用计算机通过互联网传递信息的通信手段。

7．FTP——文件传输协议。它用于在互联网上传输文件或从网络上下载文件。

8．RS-232：目前RS-232是PC与通信工业中应用最广泛的一种串行接口。RS-232被定义为一种在低速率串行通信中增加通信距离的单端标准。RS-232采取不平衡传输方式，即所谓单端通信。

9．RJ45：是一个常用名称，指的是由IEC (60)603-7标准化，使用由国际性的接插件标准定义的8个位置（8针）的模块化插孔或者插头。IEC(60)603-7也是ISO/IEC 11801国际通用综合布线标准的连接硬件的参考标准。

Exercises（练习）

1. Match the explanations in column B with words and expressions in column A.（搭配每组中同意义的词或短语）

A	B	A	B
集线器	Modem	无线局域网	NIC
环形网络	Fiber optic cabling	主机	Workstation
微波	Network protocol	工作站	E-mail
调制解调器	Microwave	广域网	WLAN
网络协议	Ring network	网卡	LAN
光缆	Website	电子邮件	Host
网络站点	Hub	局域网	WAN

2. Choose the proper words to Fill in the blanks.（选词填空）

What Is Network?

A group of two or more computer systems linked together. There are many types of computer networks, including:

(____________________): The computers are geographically close together (that is, in the same building).

(____________________): The computers are farther apart and are connected by telephone lines or radio waves.

(________________________): The computers are within a limited geographic area, such as a campus or military base.

(________________________): A data network designed for a town or city.

(________________________): A network contained within a user's home that connects person's digital devices.

In addition to these types, the following characteristics are also used to categorize different types of networks:

(________________________): The geometric arrangement of a computer system. Common topologies include bus, star, and ring. See the Network topology diagrams in the Quick Reference section of Webopedia.

(________________________): The protocol defines a common set of rules and signals that computers on the network used to communicate. One of the most popular protocols for LANs is called Ethernet. Another popular LAN protocol for PCs is the IBM token-ring network .

(________________________): Networks can be broadly classified as using either a peer-to-peer or client/server architecture.

Computers on a network are sometimes called nodes. Computers and devices that allocate resources for a network are called servers.

> Words to be chosen from:（可选词）
> home-area networks (HANs), local-area networks (LANs), protocol,
> campus-area networks (CANs), architecture , topology
> wide-area networks (WANs), metropolitan-area networks (MANs),

3．Translate the following English Sentences into Chinese.（将下面英语句子翻译为中文）

(1) Internet is a global network connecting millions of computers. More than 100 countries are linked into exchanges of data, news, and opinions.

(2) Router is a device that forwards data packets along networks. A router is connected to at least two networks, commonly two LANs or WANs or a LAN and its ISP.

(3) Routers are located at gateways, the places where two or more networks connect.

(4) World Wide Web is a system of Internet servers that support specially formatted documents.

(5) There are several applications called Web browsers that make it easy to access the World Wide Web; Two of the most popular are Netscape Navigator and Microsoft's Internet Explorer.

4．实践作业：

（1）到本校机房查看局域网的组成情况。

（2）上网申请一个自己的电子邮箱，并向同学发电子邮件，感受互联网带来的便利。

（3）利用 QQ、微信等通信工具与同学交流学习心得。

（4）利用百度、知乎、360 搜索、搜狗等搜索引擎检索有关量子计算机的信息。

附文3：Reading Material（阅读材料）

1. China's IPv6 Network to Use Juniper Technology

China's Next Generation Internet project, formed to promote IPv6 throughout China, will be using Juniper Networks' (JNPR) M- and T-series routing platforms.

Juniper Networks platforms will be deployed in China's Next Generation Internet's (CNGI) participating networks, including China Education and Research Network (CERNET2), China Mobile, China Netcom, China Railcom, China Telecom, and China Unicom.

The CNGI project was launched in 2003 by China's National Development and Reform Commission (NDRC), and its charter is to create a next-generation national IPv6 backbone covering 20 cities and 39 massive network points of presence (GigaPOPs) to provide pervasive advanced IP services.

IPv6 is the next generation Internet Protocol, and improves on the current version by greatly multiplying the number of IP addresses to accommodate the growing number of networked devices, users and applications with built-in security mechanisms.

Once deployed, the Juniper Networks platforms will provide a test bed for Multiprotocol Label Switching (MPLS) and enable the service providers to offer a wide variety of advanced IPv6-based services in the near future.

These services include Virtual Private Networking (VPN), high-quality IP voice and video streaming, as well as third-generation (3G) mobile applications. The M- and T-series platforms will also provide interoperability with networks based on the current version Internet Protocol, IPv4, by using advanced IPv4-to-IPv6 interworking features supported by the industry-leading JUNOS operating system.

These JUNOS features will also enable CNGI to facilitate the seamless transition of IPv4 networks to IPv6 as the networks expand to support the growing needs of businesses, government, institutions, and individuals in the dynamic Chinese economy.

2. Intranet

With the advancement made in browser-based software for the Internet, many private organizations are implementing intranets. An intranet is a private network utilizing Internet-type tools, but available only within that organization. For large organizations, an intranet provides an easy access mode to corporate information for employees.

3. Ethernet

Ethernet is the most popular physical layer LAN technology in use today. Other LAN types include Token Ring, Fast Ethernet, Fiber Distributed Data Interface (FDDI), Asynchronous Transfer Mode (ATM), and LocalTalk. Ethernet is popular because it strikes a good balance between speed, cost, and ease of installation. These benefits, combined with wide acceptance in the computer marketplace and the ability to support virtually all popular network protocols, make Ethernet an ideal networking technology for most computer users today. The Institute for Electrical and Electronic Engineers (IEEE) defines the Ethernet standard as IEEE Standard 802.3. This standard defines rules for configuring an Ethernet network as well as specifying how elements in an Ethernet network interact with one another. By adhering to the IEEE standard, network equipment and network protocols can communicate efficiently.

4. Fast Ethernet

For Ethernet networks that need higher transmission speeds, the Fast Ethernet standard (IEEE 802.3u) has been established. This standard raises the Ethernet speed limit from 10 Megabits per second (Mbps) to 100 Mbps with only minimal changes to the existing cable structure. There are three types of Fast Ethernet: 100BASE-TX for use with level 5 UTP cable, 100BASE-FX for use with fiber-optic cable, and 100BASE-T4 which utilizes an extra two wires for use with level 3 UTP cable. The 100BASE-TX standard has become the most popular due to its close compatibility with the 10BASE-T Ethernet standard. For the network manager, the incorporation of Fast Ethernet into an existing configuration presents a host of decisions. Managers must determine the number of users in each site on the network that need the higher throughput, decide which segments of the backbone need to be reconfigured specifically for 100BASE-T and then choose the necessary hardware to connect the 100BASE-T segments with existing 10BASE-T segments. Gigabit Ethernet is a future technology that promises a migration path beyond Fast Ethernet so the next generation of networks will support even higher data transfer speeds.

5. Token Ring

Token Ring is another form of network configuration which differs from Ethernet in that all messages are transferred in a unidirectional manner along the ring at all times. Data is transmitted in tokens, which are passed along the ring and viewed by each device. When a device sees a message addressed to it, that device copies the message and then marks that message as being read. As the message makes its way along the ring, it eventually gets back to the sender who now notes that the message was received by the intended device. The sender can then remove the message and free that token for use by others.

Various PC vendors have been proponents of Token Ring networks at different times and thus these types of networks have been implemented in many organizations.

6. Protocols

Network protocols are standards that allow computers to communicate. A protocol defines how computers identify one another on a network, the form that the data should take in transit, and how this information is processed once it reaches its final destination. Protocols also define procedures for handling lost or damaged transmissions or "packets". TCP/IP (for UNIX, Windows NT, Windows 95, and other platforms), IPX (for Novell NetWare), DECnet (for networking Digital Equipment Corp. computers), AppleTalk (for Macintosh computers), and NetBIOS/NetBEUI (for LAN Manager and Windows NT networks) are the main types of network protocols in use today.

Although each network protocol is different, they all share the same physical cabling. This common method of accessing the physical network allows multiple protocols to peacefully coexist over the network media, and allows the builder of a network to use common hardware for a variety of protocols. This concept is known as "protocol independence", which means that devices that are compatible at the physical and data link layers allow the user to run many different protocols over the same medium.

7. Collisions

Ethernet is a shared medium, so there are rules for sending packets of data to avoid conflicts and protect data integrity. Nodes determine when the network is available for sending packets. It is possible that two nodes at different locations attempt to send data at the same time. When both PCs are transferring a packet to the network at the same time, a collision will result.

Minimizing collisions is a crucial element in the design and operation of networks. Increased collisions are often the result of too many users on the network, which results in a lot of contention for network bandwidth. This can slow the performance of the network from the user's point of view. Segmenting the network, where a network is divided into different pieces joined together logically with a bridge or switch, is one way of reducing an overcrowded network.

Chapter 4 Software（软件）

教学要求

掌握专业关键词汇（key words）；能阅读本章所列英语短文；了解各种操作系统软件和应用软件。

教学内容

各种操作系统；办公软件、图像处理软件；常用的专业术语。

教学提示

通过在某种操作系统环境下使用各种应用软件，熟悉本章内容，尤其是办公软件和图像处理软件。

4.1 Operating System（操作系统）

So long as we turn on the computer, it starts to run the procedure. The first procedure a computer runs is the **operating system**. Why does a computer firstly run the operating system, but not application procedure, for example, the Word? The operating system is the "intermediate" of application procedure and computer hardware. If there is not the unified arrangement and management of the operating system, the computer hardware cannot carry out the application procedure at all. The operating system provides an interactive interface for the computer hardware and the application procedure, and it is also in charge of the basic work of each part of computer hardware. Normally, **functions** relevant to the operating system are: **job management**; **resource management**; control of I/O operations; **error recovery; memory management**.

只要我们打开计算机，计算机就开始运行程序。计算机运行的第一个程序就是操作系统。为什么首先运行操作系统，而不直接运行像 Word 这样的应用程序呢？操作系统是应用程序与计算机硬件的“媒介”。如果没有操作系统的统一安排和管理，计算机硬件就根本没有办法执行应用程序的命令。操作系统为计算机硬件和应用程序提供了一个交互的界面，指挥计算机各部分硬件的基本工作。与操作系统相关的功能通常有：作业管理、资源管理、I/O 操作控制、差错排除和存储管理。

 **Key words:** operating system（操作系统）, function（功能）, job management（作业管理）, resource management（资源管理）, error recovery（差错排除）, memory management（存储管理）

常见操作系统有：Windows 7、Linux、Windows 8、Windows 10 等。

朗读音频

Operating System

Windows 7

Windows 7 is the operating system developed by Microsoft, and kernel version number is Windows NT 6.1. Windows 7 is available for home and commercial work environment, Notebook computer, Tablet computer, multimedia center, etc. In July 14, 2009 Windows 7 RTM (Build 7600. 16385) officially launched, and on October 22, 2009 in the United States, Microsoft released Windows 7. Windows 7 also released a version of Windows Server 2008 R2. The early hours of February 23, 2011 , Microsoft for mass users of the official release of Windows 7 upgrade patch—Windows 7 SP1（Build 7601.17514. 101119-1850）, also including Windows Server 2008 R2 SP1 upgrade patch.

Windows 7 是由微软公司（Microsoft）开发的操作系统，核心版本号为 Windows NT 6.1。Windows 7 可供家庭及商业工作环境、笔记本电脑、平板电脑、多媒体中心等使用。2009 年 7 月 14 日，Windows 7 RTM（Build 7600.16385）正式上线，2009 年 10 月 22 日，微软于美国正式发布 Windows 7。Windows 7 同时还发布了服务器版本——Windows Server 2008 R2。2011 年 2 月 23 日凌晨，微软面向大众用户正式发布了 Windows 7 升级补丁——Windows 7 SP1（Build 7601.17514.101119-1850），另外还包括 Windows Server 2008 R2 SP1 升级补丁。

Windows XP

Windows XP is the next version of Microsoft Windows following Windows 2000 and Windows Millennium. Windows XP brings the convergence of Windows operating system by integrating the strengths of Windows 2000 — standards–based security, manageability, and reliability with the best features of Windows 98 and Windows Me—Plug and Play, easy-to-use user interface, and innovative support services to create the best Windows yet.

Windows XP 是继 Windows 2000 与 Windows Me 之后的下一个 Microsoft Windows 版本。Windows 2000 的强大优势体现为基于标准的安全性、可靠性及管理功能；而 Windows 98 与 Windows Me 的最佳特性则以即插即用功能、简易用户界面及创新支持服务为代表。Windows XP 正是集上述 Windows 操作系统之大成而创造出的空前优秀的 Windows 产品。

Intelligent User Interface

智能化用户界面

Fast User Switching for Multiple User of a Computer

多用户的计算机上的快速用户切换

New Visual Style

新的视觉类型

Redesigned Start Menu

重新设计的“开始”菜单

Search Companion

搜索伴侣

My Documents

我的文档

Webview

Webview

File Grouping

文件分组

User Interface Enhances Productivity

用户界面提高了办公效率

Notes

1. Plug and Play: 即插即用是微软为其 Win95 和后续操作系统开发的一种能力，该能力使计算机能自动识别设备。

2. Webview: Windows XP 使用了 Webview 技术来帮助您更好地管理文件和文件命名空间。比如说：如果选择了一个文件或文件夹，您会看见一个选项列表，这个列表允许您对文件进行重命名、移动、复制，或将文件作为 E-mail 发送或发布到 Web 上。

Linux

Linux runs on Intel 80x86-based machines, where x is 3 or higher. Linux is also very portable and flexible because it has now been ported to DEC Alpha, PowerPC, and even Macintosh machines. And progress is being made daily by Linux **enthusiasts** all

Linux 运行在基于 Intel 80x86 的机器上，这里的 x 是指 3 或比 3 大的数字。Linux 非常易于移植，灵活，所以现在已经移植到如 DEC Alpha、Power PC 甚至 Mac 机器上。全世界所有的 Linux

over the world to make this free operating system available to all the popular computing machines in use today. Because the source code for the entire Linux operating system is freely available, developers can spend time actually porting the code instead of wondering or worrying about whom to pay benefit **licensing** fees.

爱好者每天都在取得进展，使这个免费的操作系统能运行在当今所有的普及型机器上。因为整个Linux操作系统是免费的，所以开发者实际上只需花费时间来移植代码，却不必担心要向谁支付受益许可费。

Notes

1. be ported to...: 被移植到……
2. in use today: 定语，修饰 machine

Key words: Linux（一种自由操作系统），enthusiast（狂热者），license（许可证）

Windows 8

Windows 8 is a new generation of operating system after Windows 7 which was officially launched by Microsoft on 26 October 2012.The Windows 8 operating system is a revolution change which has the unique start interface and,touch interaction system and aim to make people's daily operation more easily and quickly, Windows 8 provides efficient and **convenient** work **environment**. It supports chip **architecture** from Intel, AMD, and ARM, which was made by Microsoft research Cambridge and ETH Zurich. The system has a better durability, it starts faster and uses less memory, and it was **compatible** with Windows 7 supported software and hardware.

Windows 8是继Windows 7之后的新一代操作系统，是由Microsoft公司于2012年10月26日正式推出的。Windows 8操作系统具有革命性的变化，系统独特的开始界面和触控式交互系统，旨在让人们的日常操作更加简单和快捷，为人们提供高效、方便的工作环境。它支持来自Intel、AMD和ARM的芯片架构，由微软剑桥研究院和苏黎世联邦理工学院联合开发。该系统具有更好的续航能力，且启动速度更快、占用内存更少，并兼容Windows 7所支持的软件和硬件。

Notes

aim to：目的是，目的在于

Key words: convenient（方便的），environment（环境），architecture（架构），compatible（兼容的）

Windows 10

Windows 10 is a new generation of **cross-platform** operating system which is researched and developed by Microsoft. All **eligible** Windows 7, Windows 8.1 and Windows Phone 8.1 users will be able to **upgrade** to the Windows 10 for free. Microsoft will provide a permanent life cycle support for all equipment which has been upgraded to the Windows 10. The Windows 10 operating system restore the **original** start menu, so that it's easier to old customers to use Windows 10 desktop. Windows 10 may be the last Windows version of Microsoft published, and the next generation of Windows will appear as the Update form.

Windows 10 是美国微软公司所研发的新一代跨平台操作系统。所有符合条件的 Windows 7、Windows 8.1 以及 Windows Phone 8.1 用户都将可以免费升级到 Windows 10。所有升级到 Windows 10 的设备，微软都将提供永久生命周期的支持。Windows 10 操作系统恢复原来的开始菜单，以便老用户更容易使用 Windows 10 的桌面。Windows 10 可能是微软发布的最后一个 Windows 版本，下一代 Windows 将以升级形式出现。

Notes

1. be able to：能够
2. appear as：作为……出现
3. so that：以便，所以

Key words: cross-platform（跨平台的），eligible（合适的、符合条件的），upgrade（升级），original（原来的）

4.2 Office Software（办公软件）

Microsoft Office 2010

Microsoft Office 2010 is Microsoft's office software. Office 2010 can support 32-bit and 64-bit Vista and Windows 7. But it only supports 32-bit of the Windows XP. Office 2010 provides four versions. They are the Office 2010 Small Business Edition, Office 2010 Professional Edition, Office 2010 Professional Plus Edition, and Office Web Apps.

Office 2010 Small Business Edition has three features. It can flexibly and efficiently manage the business. E-mail experience is a comprehensive upgrade. Daily affairs perfect order. It includes Word, Excel, PowerPoint, Outlook, and OneNote. Office 2010 Professional Edition also has three features. It can enhance the insight and increase profits. Let us to easily communicate with customers. It has a high-performance tool. It contains Word, Excel, PowerPoint, Outlook, OneNote, Access, and Publisher. Office 2010 Professional Plus version, we can use it anytime, anywhere. This will allow our ideas become a reality. It contains Word, Excel, PowerPoint, Outlook, OneNote, Access, Publisher, InfoPath, Share-Point Workspace, and Communicator. Office Web Apps allows you to more easily carry Office.

Microsoft Office 2010 是微软推出的办公软件。Office 2010 可支持 32 位和 64 位 Vista 及 Windows 7。但是它仅支持 32 位 Windows XP。Office 2010 有 4 个版本，它们是 Office 2010 小型企业版、Office 2010 专业版、Office 2010 专业增强版和 Office Web Apps。

Office 2010 小型企业版有 3 个特点：它可以灵活高效地管理业务；电邮体验全面升级；它让日常事务井井有条。它包含 Word、Excel、PowerPoint、Outlook 和 OneNote。Office 2010 专业版也有 3 个特点：它可以提升洞察力，增加利润；让我们轻松与客户保持沟通；它有高性能工具。它包含 Word、Excel、PowerPoint、Outlook、OneNote、Access 和 Publisher。而至于 Office 2010 专业增强版，我们可以随时随地使用它，这样能让我们的创意成为现实。它包含 Word、Excel、PowerPoint、Outlook、OneNote、Access、Publisher、InfoPath、SharePoint Workspace 和 Communicator。Office Web Apps 让您在任何地方都能使用 Office。

Microsoft Office 2016

Microsoft Office 2016 is Microsoft's latest generation of office software, including **collaboration** tools and cloud support which are the biggest improvement in all 30 years of Office history.

Microsoft Office 2016 是微软推出的最新一代办公软件，其中的协作工具和云端支持都是 Office 30 年历史上的最大改进。

Office 2016 mainly has the following versions: Office 365, Office 2016 Professional Edition, Office 2016 Home and Student Edition, Small Business Edition, Project Professional Edition, and Visio Standard Edition and Professional Edition. The main contents are as follows:

Office 2016 主要有以下版本：Office 365、Office 2016 专业版、Office 2016 家庭和学生版、小型企业版、Project 专业版、Visio 标准版和专业版等。主要版本内容如下。

Office 2016 Suit (Office 365, Office 2016 Professional Edition, Office 2016 Home and Student Edition, Small Business Edition) is the collection tools of powerful office software of Microsoft Office 2016, including Word, Excel, PowerPoint, OneNote, Outlook, Skype , Project, Visio, and Publisher, and so on.

Office 2016 套装类（Office 365、Office 2016 专业版、Office 2016 家庭和学生版、小型企业版）是 Microsoft Office 2016 强大办公软件的集合工具，其中包括 Word、Excel、PowerPoint、OneNote、Outlook 、Skype 、Project 、Visio 以及 Publisher 等组件和服务。

Office 2016 Project is a member of the Office 2016 **component,** which is mainly used for the management of projects and tasks, and it can effectively strengthen the ability of collaborative work.

Office 2016 Project 是 Office 2016 办公组件中的一个，主要用于对项目和任务的管理，可以有效地加强协同工作的能力。

Office 2016 Visio can help us to create **professional-looking** charts, in order to understand, record, and analyze information, data, system, and process.

Office 2016 Visio 能够帮助我们创建具有专业外观的图表，以便理解、记录和分析信息、数据、系统和过程。

 **Key words:** collaboration（协作），component（组件），professional-looking（专业外观的）

4.3 Graphics Software（图像软件）

4.3.1 Photoshop（图像专家）

Adobe Photoshop

Adobe Photoshop software delivers even more imaging magic, new creative options, and the Adobe Mercury Graphics Engine for blazingly fast performance. Retouching with new Content-Aware features, it can create superior designs as well as movies using new and reimagined tools and workflows.

Adobe Photoshop软件具备更为先进的图像处理技术和全新的创意选项，且使用了全新的Adobe Mercury图像引擎，拥有前所未有的性能和极快的处理速度。借助新增的“内容识别”功能进行润色并使用全新和改良的工具和工作流程，它能创建出出色的设计和影片。

What's the features in Adobe Photoshop?

Adobe Photoshop有哪些功能呢?

Enhance your creativity and boost your productivity with the new Adobe Mercury Graphics Engine, groundbreaking new Content-Aware tools, reengineered design tools, and more.

使用全新的Adobe Mercury图形引擎、突破性的全新“内容识别”工具、经过改良的设计工具等，来增强你的创意，提高你的工作效率。

Content-Aware features take image retouching to a new level of ease and control. Get superior results when you crop, correct wide-angle lens curvatures, auto-correct, and more.

“内容识别”功能，让你更轻松精确地控制图像润色。在你完成裁剪、纠正广角镜头曲率、自动纠正等更多其他任务时能够获得更好的效果。

Edit at blazingly fast speeds with the Adobe Mercury Graphics Engine. Boost your productivity with new preset migration and sharing, new Background Save and auto-recovery options, and a modern user interface.

Adobe Mercury图形引擎让你能够以惊人的速度编辑。全新的预设转移和共享、自动恢复和后台存储选项以及现代化的用户界面能够大大提高你的工作效率。

Adobe Photoshop mainly has the following versions: CS, CS2, CS3, CS4, CS5, CS6, CC, CC 2014, CC 2015, CC 2017.

Adobe Photoshop主要有以下版本：CS、CS2、CS3、CS4、CS5、CS6、CC、CC 2014、CC 2015、CC 2017。

4.3.2 CorelDRAW（平面设计）

CorelDRAW

CorelDRAW is Corel Software's products in Canada. It is a vector-based graphics and layout software. It is used in many fields. For example, the logo designing, illustration painting, typography, and so on.

CorelDRAW 是加拿大 Corel 软件公司的产品，它是一个基于矢量图的绘图与排版软件。它应用于诸多领域，例如，商标设计、插图绘画、排版等。

The software consists of two graphics applications. Its main feature is a vector illustration and page design and image editing. Set of mapping software combination gives users a powerful interactive tool. With simple operation, it allows the user to create a variety of rich and dynamic special effects and immediate results of bitmap images. CorelDRAW full range of design and web page functionality allows full flexibility of design. The software package for professional designers and graphics enthusiasts offers presentations, color pages, brochures, product packaging, logos, web pages, and other. The software provides intelligent drawing tools and a new dynamic wizard. It can substantially reduce the difficulty of manipulation of the user. It reduces the click step and save design time.

该软件包含两个绘图应用程序。它的主要功能是矢量图及页面设计和图像编辑。这套绘图软件组合带给用户强大的交互式工具。在简单的操作中，它使用户可创作出多种富于动感的特殊效果及点阵图像即时效果。CorelDRAW 全方位的设计及网页功能可以让设计方案充满灵活性。该软件套装为专业设计师及绘图爱好者提供简报、彩页、手册、产品包装、标识、网页等。该软件提供了智慧型绘图工具以及新的动态向导。它可以充分降低用户的操控难度、减少点击步骤和节省设计时间。

4.3.3 CAD（计算机辅助设计）

AutoCAD

AutoCAD is Autodesk Corporation's leading product. Now it is the most popular two-dimensional **plot** software, and it has the broad user group in the two-dimensional plot domain. AutoCAD has the **formidable**

AutoCAD 是 Autodesk 公司的主导产品，是当今最流行的二维绘图软件，它在二维绘图领域拥有广泛的用户群。AutoCAD 有强大的二

two-dimensional function, such as plot, edition, **profile** line and design plan, mark note as well as the second developments. **Simultaneously**, it has partial three-dimensional function. AutoCAD provides AutoLisp, ADS, and ARX which are used as the second development tool. In many practical applications domains (for example machinery, architecture, electron), some software developers have developed much practical application software on the bases of AutoCAD.

维功能，如绘图、编辑、剖面线和图案绘制、尺寸标注以及二次开发等，同时它还有部分三维功能。AutoCAD 提供 AutoLISP、ADS 和 ARX 作为二次开发的工具。在许多实际应用领域（如机械、建筑、电子）中，一些软件开发商已在 AutoCAD 的基础上开发出许多符合实际应用的软件。

Key words: plot[（建筑物等的）平面图；标绘图], formidable（强大的；杰出）, two-dimensional（二维的）, profile（剖面）, simultaneously（同时）

4.3.4 3ds Max（建模圣手）

3ds Max

3ds Max runs in Windows **platform**. A complete **multithreading** formidable software, which is fully able to use the **multi-processor** and **render** any network. The software has been famous for its **integrated** and **intellectualized** interface since its birth. The integration refers to that all work, such as the three dimensional modeling, the two-dimensional **lofting**, the frame editor, the material quality edition, the animation setting, which are all completed in a **unified** interface, avoiding the trouble of screen switch. The so-called intellectualization is that unless the current operation conforms to you to make some animation, it could be permitted to continue, contrarily, some orders cannot be used by you.

3ds Max 运行于 Windows 平台。它是一个完全多线程，可充分发挥对称多处理器的作用，并在任意网络条件下都能进行渲染的强大软件。这个软件自诞生以来，就以一体化、智能化界面著称。一体化是指所有工作，如三维造型、二维放样、帧编辑、材质编辑、动画设置等都在统一的界面中完成，这样就避免了屏幕切换带来的麻烦。所谓智能化就是能在当前符合你制作某动画时能被用上的命令才能被你所用，反之，某些命令就不能被使用。

The 3ds Max 2017 software adds the following new features: viewport zoom is now infinite, mirror tool now has geometry mode, offset controllers are added, a lot of new tools are added in curve editor and buffer curves, UV editor can be flatten by material id and smoothing groups, and so on.

3ds Max 2017 软件增添了以下新功能：视口缩放现在是无限的，镜像工具现在有几何体模式，增加了偏移控制器，曲线编辑器和缓冲区曲线中添加了许多新工具，UV 编辑可通过材质 ID 和光滑组来展平等。

3ds Max 的 5 个功能模块：①建模（modeling object）；②材质设计（material design）；③灯光和相机（lighting and camera）；④动画（animate）；⑤渲染（rendering）。

Key words: platform（平台），multithreading（多线程），multi-processor（多处理器），render（渲染），integrated（集成的），intellectualized（智能化的），loft（放样），unified（统一的）

Notes

操作命令：选择（selection）、移动（migration）、缩放（reproduce）、镜像（mirror）、阵列（array）、视图（view）。

二维图案的编辑通过"edit spline"来实现。

4.3.5 Flash（平面动画）

Adobe Flash

Flash movies are graphic and animation for Web sites. They consist primarily of **vector graphics**, but they can also contain imported **bitmap** graphics and sounds. Flash movies can **incorporate interactivity** to permit input from viewers, and you can create **nonlinear** movies that can interact with other Web applications. Web designers use Flash to create **navigation** controls system, **animated logos**, long-form animations with **synchronize**d sound, and even colourful **sensory**-rich Web sites. Flash movies are **compacted-vector** graphics, so they are downloaded rapidly, and the scale is flexible depending on the viewer's screen size.

Flash 动画就是网站上的图形和动画。它主要是由矢量图形构成的，也可包含位图和声音。Flash 动画具有互动性，动画观赏者可对影片进行输入操作。Flash 还可创作出非线性的动画，这种动画可以与别的网络应用软件交互使用。网页设计者可用 Flash 创建导航控制系统、动画商标字、与声音同步的大型动画，甚至还可以创建感官体验丰富的网站。由于 Flash 动画是经压缩的矢量图形，因此下载速度快，其大小比例还可视观看者计算机的屏幕大小而定。

Flash 小知识

矢量图：例如一个圆，矢量图要记录圆心坐标、半径、线条粗细和填充色彩等。文件体积很小，也可以很容易地进行放大、缩小或旋转等操作，并且不会失真。

点阵图：由许多点组成，这些点称为像素，不同色彩的像素合在一起就构成了一幅完整的图像。在保存点阵图时要记录的是每一个像素的位置和色彩数据，所以比矢量图大得多。

You've probably watched and interacted with Flash movies on many Web sites, including **Disney**, the Simpson's, and **Coca-Cola**. Millions of Web users have received the Flash Player with their computers, browsers, or system software; others have downloaded it from the Macromedia Web site. The Flash Player resides on the local computer, where it plays back movies in browsers or as stand-alone applications .Viewing a Flash movie on the Flash Player is similar to viewing a video tape on a **VCR**. The Flash Player is the device used to display the movies you create in the Flash authoring application.

或许你在许多网站，如迪士尼的网站、Simpson 的网站、可口可乐的网站等，观看过或交互使用过 Flash 动画。成千上万的网友通过计算机、浏览器或系统软件接收过 Flash 播放器，有的可能是从 Macromedia 的网站上下载的。Flash 播放器一般置于本地计算机上。在那里，Flash 播放器以浏览器的形式播放动画，或像独立的应用软件一样播放动画。在 Flash 播放器上观看动画就如同在录像机上观看录像带。Flash 播放器是一种显示你用 Flash 创作应用软件创作的动画的装置。

Flash 动画的几种方式

第一种是简单的逐帧动画，和制作动画电影一样，一幅一幅地画，然后按照顺序连续播放，最形象的例子就是倒计时的动画。

第二种是 Shape 动画，也就是形状改变的动画。这种改变在 Flash 中受人控制的程度较小，主要是依靠计算机自己运算，这样形成的动画千奇百怪，而且极可能各不相同，例如字母之间的形状变化。

第三种是 Motion 动画，也就是移动动画。它是指一个物体发生空间及大小上的变化，例如文字的飘动。

Key words: vector graphics（矢量图形），bitmap（位图），incorporate（合并，融入），interactivity（互动性，交互性），nonlinear（非线性），navigation（航海），animated（活泼的，动画的），logo（logogram 的简写，标志），synchronize（使同步，同时发生），sensory（知觉的，感官的），compact-vector（压缩向量），Disney（迪士尼），Coca-Cola（可口可乐），VCR（video cassette recorder，录像机）

4.4 Tool Software：Ghost（工具软件：Ghost）

Ghost functions：Save the **stuff** you care about. Automatically back up and recover everything on your computer.

Ghost 的作用：保存你的重要资料，自动备份并恢复到你的计算机上。

Tool Software: Ghost

4.4.1 Key Features（主要特征）

Backup everything on your computer—digital music, photos, financial documents, applications, settings, operating system, etc. in one easy step.

只需一个简单步骤就可备份你计算机上的所有数据，包括数字音乐、照片、金融单据/财务文件、应用软件、配置、操作系统等。

Recovers your system and data even when you can't restart your operating system.

即使在你的操作系统不能启动的情况下也可以恢复系统及其数据。

Makes incremental backups to maximize space and save time.

利用增量备份使空间最大化并节约时间。

Makes backups on the fly, without restarting your system.

不需要重启系统而直接备份。

Backs up to almost any media, including CDR/RW and DVD+-R/RW drives, USB and FireWire® (IEEE 1394) devices, and Iomega® Zip® and Jaz® drives.

几乎可以备份到任何介质，包括CD/RW和DVD+-R/RW驱动器，USB和1394设备、Iomega ZIP软盘和Jaz驱动器等。

Key Features

4.4.2 Latest Features（最新特征）

Automatically creates an initial backup **schedule** based on your computer's configuration.

自动根据计算机配置创建初始备份任务。

Automatically detects storage devices, analyzes your system, and offers practical backup advice during installation.

自动检测存储介质，自动分析系统，并在安装过程中提供最佳备份建议。

Automatically monitors and optimizes backup disk space.

自动监视和优化备份磁盘空间。

Triggers backups on key events, like new program installations or user logins.

Creates new backups on demand with One Button "Back up Now".

Encrypts backups to help keep them secure.

Task-based interface simplifies management and monitoring.

Displays all scheduled backups—plus the degree of backup protection for each drive on your computer—in one convenient view.

在关键事件如安装新程序或用户登录时自动触发备份。

通过"一键备份"按钮创建新备份。

对备份加密以保证数据安全。

基于任务的接口可以简化管理和监视。

用简易方便的视图模式可以显示所有的备份任务，甚至对计算机的每个驱动器进行分级保护。

4.4.3 Partion Backup and Image It（磁盘分区备份为映像文件）

Main Steps:

1. In DOS mode enter command:

D:\ghost>GHOST

2. Start

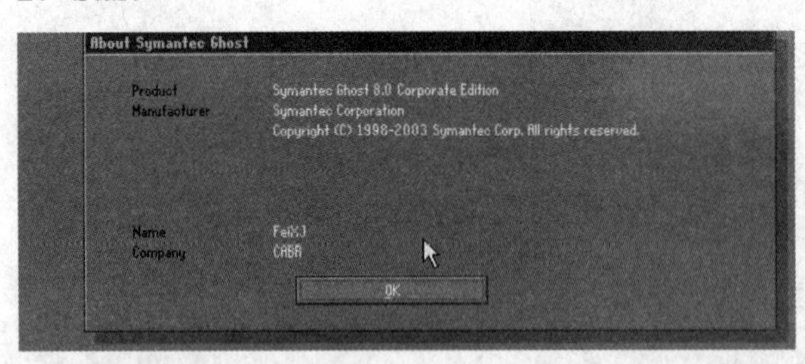

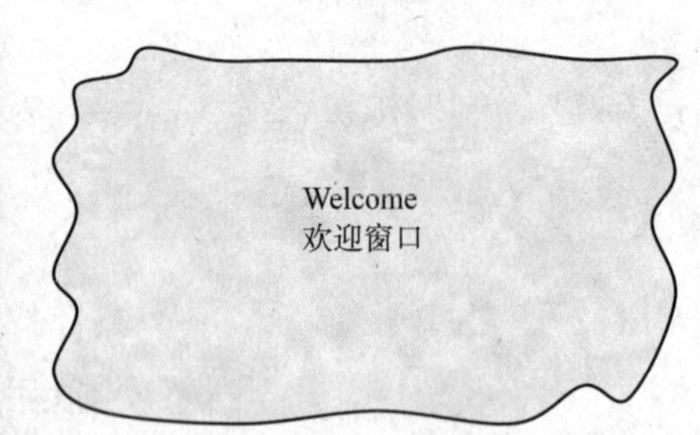

3. Partition → To Image

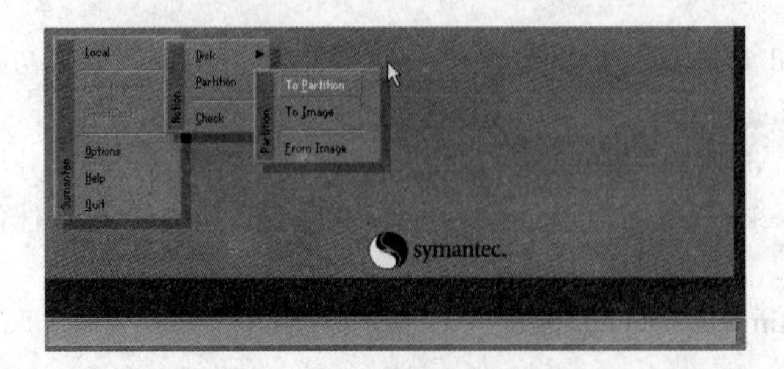

Partition → To Image
把指定分区所有信息备份到指定磁盘（指定磁盘容量要足够）

4. Select local source drive

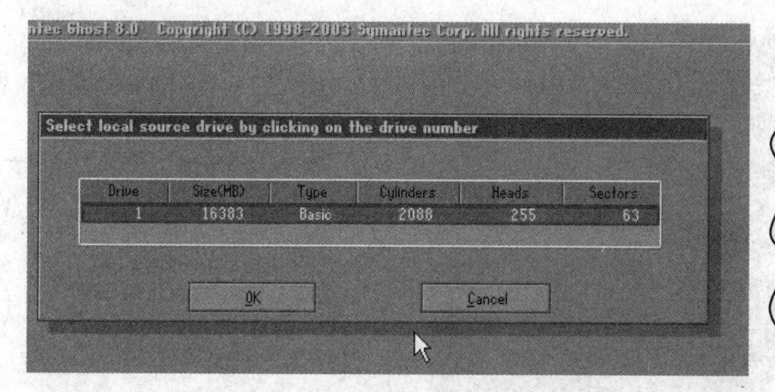

Select local source drive
选择要备份的本地磁盘驱动器

5. Select source **partition**(s) from basic drive

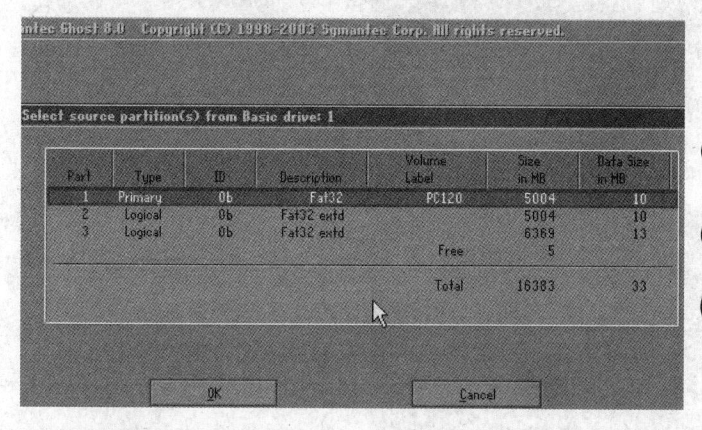

Select source partition (s) from basic drive
从驱动器上选择分区

6. Enter image file name to restore from image

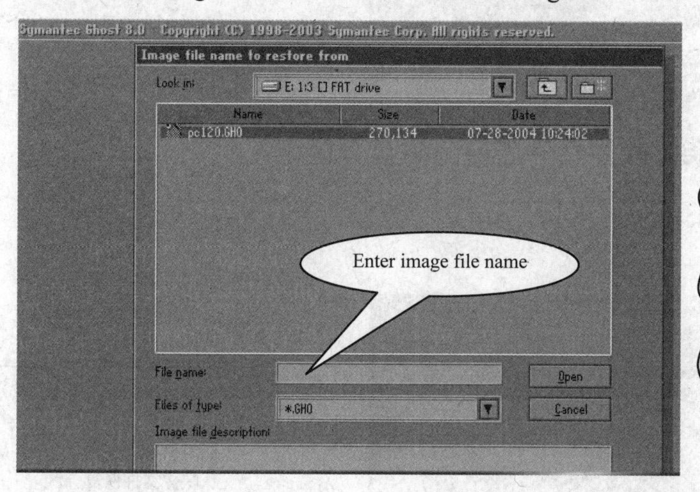

Enter image file name to restore from image
输入用于恢复的映像文件

7. Compress image file

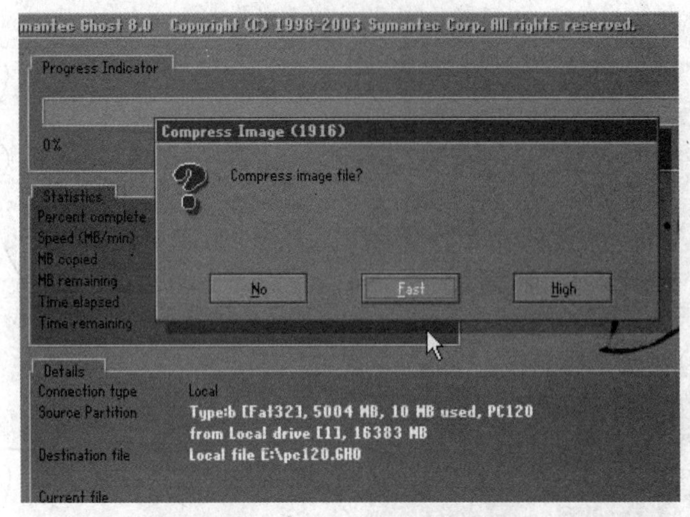

Compress image file
选择映像文件的压缩方式（不压缩、快速或高压缩）

8. Image creation completed successfully

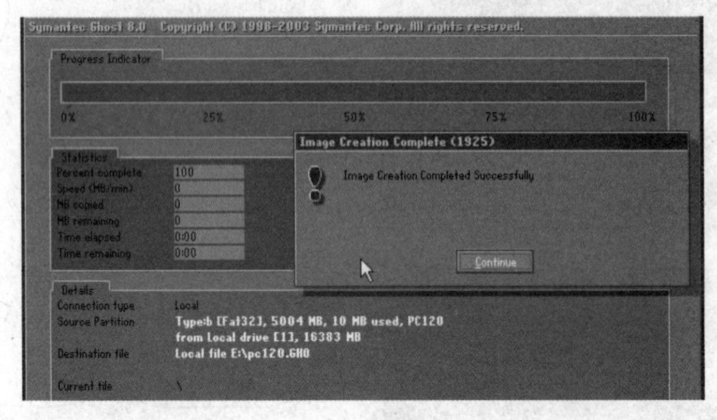

Image creation completed successfully
映像文件创建完成

4.4.4 Partition from image（映像文件还原为磁盘分区）

Main steps are similar to 4.4.3

1. Partition from image

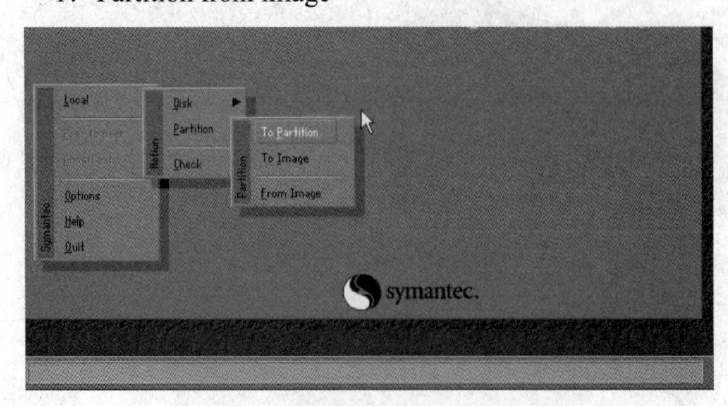

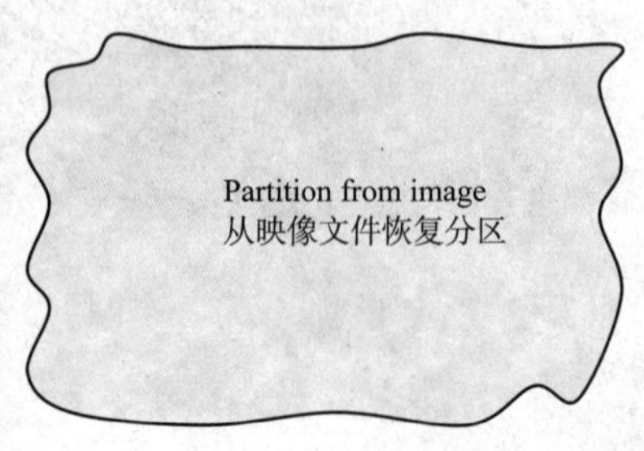

2．Select destination partition from basic drive

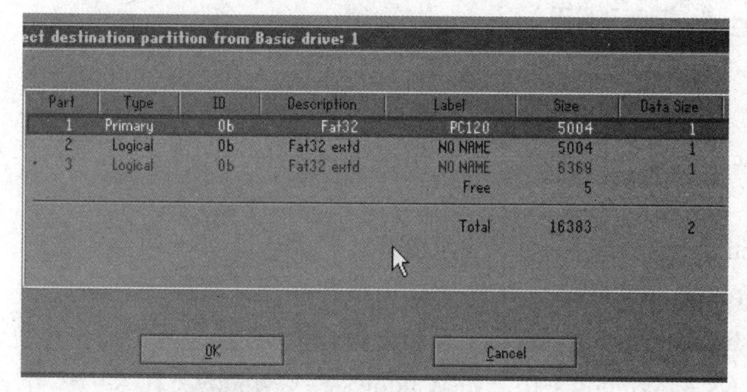

3．Image file name to restore from

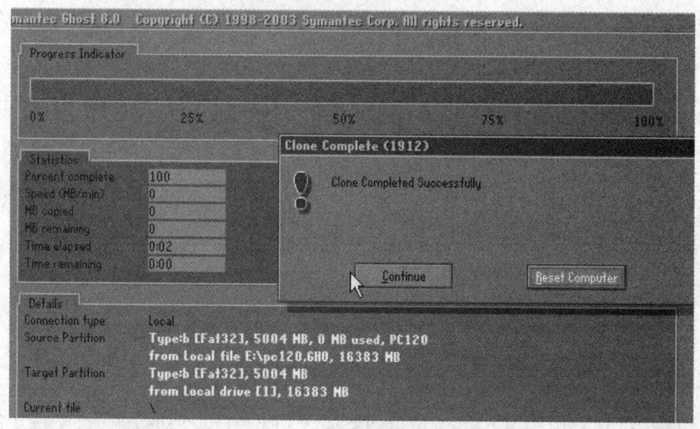

4．Restore progress

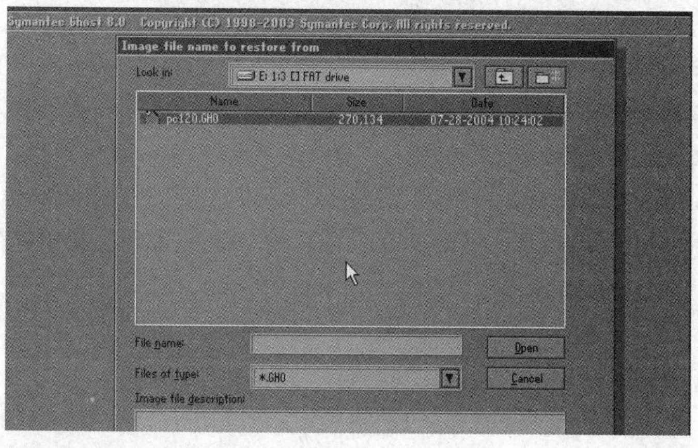

常见克隆备份工具：Windows 系统备份/恢复、diskgenius、clonezilla、杀毒软件提供的备份还原工具等。

Key words: stuff（资料），backup（备份），recover（恢复，还原），schedule（任务），trigger（触发），image（映像），encrypt（加密），partition（分区）

4.5 Situation Dialogue（情境对话）

Software Family

Shop assistant: What software do you want, sir?

Customer: I bought a computer yesterday, and I don't know what kind of software I should use?

Shop assistant: Then you should first buy the operating system, such as Windows 8, Windows10, Neoshine Linux, Deepin Linux, etc.

Customer: I mainly use it in office and drawing. What other software do I need?

Shop assistant: As for the popular office software, first you can choose Office, then the domestic WPS; there are many graphic software, such as Photoshop, CorelDraw, AutoCAD, 3ds Max, Flash, etc.

Customer: I would like to see films on computer, then what is better for that?

Shop assistant: Now the popular software is Real Player and Windows Media Player. It can play VCD. You can also see films online directly, and it's quite available.

Customer: Do I need any other necessary software besides those?

Shop assistant: Anti-virus software is necessary, such as 360 Anti-virus, Kingsoft Anti- virus, Tencent PC Manager, and the foreign Norton Anti-virus, etc.

Customer: My child likes playing games, do you have any game software?

Shop assistant: So many. You can also download the latest game online.

Customer: Ok. I'll buy all you've recommended to me.

软件家族大观园

销售员：先生，您需要什么软件？

顾客：我昨天刚买了一台计算机，还不知道该装些什么软件才好。

销售员：那您首先应该买操作系统，如 Windows 8、Windows 10、中标普华 Linux、深度 Linux 等。

顾客：我的计算机主要用于办公和作图这些方面，请问还需哪些软件？

销售员：流行的办公软件首选 Office，其次是国产的 WPS；图像软件有很多，主要有 Photoshop、CorelDraw、AutoCAD、3ds Max、Flash 等。

顾客：我还想在计算机上看电影，有什么软件比较好？

销售员：现在流行的播放软件有 Real Player 和 Windows Media Player，可播放 VCD，也可直接在线看电影，很方便。

顾客：其他还需要些什么必备的软件呢？

销售员：杀毒软件是必需的，如国产 360 杀毒软件、金山毒霸、腾讯电脑管家，国外的 Norton 杀毒软件等。

顾客：我小孩喜欢玩游戏，你这里有游戏软件吗？

销售员：游戏软件很多，最新的游戏版本可到网上直接下载。

顾客：好的，那我就把你介绍的这些软件都买上。

4.6 Reading and Compacting（对照阅读）

The GNU Operating System

The GNU Project was launched in 1984 to develop a complete UNIX like operating system which is free software: the GNU system (GNU is a recursive acronym for "GNU's Not UNIX"); it is pronounced guh-noo, like canoe. Variants of the GNU operating system, which use the kernel Linux, are now widely used; though these systems are often referred to as "Linux", they are more accurately called GNU/Linux systems.

The Free Software Foundation (FSF) is the principal organizational sponsor of the GNU Project. FSF receives very little funding from corporations or grant-making foundations, but rely on support from individuals like you.

We support the FSF's mission to preserve, protect, and promote the freedom to use, study, copy, modify, and redistribute computer software, and to defend the rights of Free Software users. We support the freedoms of speech, press, and association on the Internet, the right to use encryption software for private communication, and the right to write software unimpeded by private monopolies.

GNU 操作系统

GNU 工程开始于 1984 年，它的目标是开发一个完整的类似 UNIX 操作系统的免费软件：GNU 系统（GNU 是递归的"GNU's Not UNIX"的首字母缩写），它的发音为 guh-noo，像 canoe 的音。现在，许多的 GNU 操作系统都广泛使用的是 Linux 内核；尽管这些系统通常被称为"Linux"，但是更准确的叫法应该是"GNU/Linux"系统。

自由软件基金会（FSF）就是 GNU 工程的主要组织者和发起者。自由软件基金会从公司或发展基金会只能得到很少的资金，主要依靠个人（比如你）的支持。

我们支持自由软件基金会的任务就是保持、保护和促进自由地使用、研究、复制、修改和重新发布软件，保护自由软件用户的权利。我们支持互联网上自由的言论、评论和社团，支持私人通信使用加密软件的权利，支持不受私营垄断者影响的软件开发权利。

We Must Talk about Freedom

Estimating that today there are millions and millions users of GNU/Linux systems such as Debian GNU/Linux and Red Hat Linux. Free software has developed such practical advantages that users are flocking to it for purely practical reasons.

The good consequences of this are evident: more interest in developing free software, more customers for free software businesses, and more ability to encourage companies to develop commercial free software instead of proprietary software products.

But interest in the software is growing faster than awareness of the philosophy it is based on, and this leads to trouble. Our ability to meet the challenges and threats described above depends on the will to stand firmly for freedom. To make sure our community has this will, we need to spread the idea to the new users as they come into the community.

But we are failing to do so: the efforts to attract new users into our community are far outstripping the efforts to teach them the civics of our community. We need to do both, and we need to keep the two efforts in balance.

"Open Source"

Teaching new users about freedom became more difficult in 1998 when a part of the community decided to replace the term "free software" with "open source software".

Some who favored this term aimed to avoid the confusion of "free" with "gratis"—a valid goal. Others, however, aimed to set aside the spirit of principle that had motivated the free software movement and the GNU project, and to appeal instead to executives and business users, many of whom hold an ideology that places profit above freedom, above community, above

我们必须谈论自由

估计当今有数百万的用户使用诸如 Debian GNU/Linux 和 Red Hat Linux 的 GNU/Linux 系统。自由软件已经发挥出了这样实用的优势，使得用户纯粹为了实用原因而聚集到它身边。

这种现象的好结果是明显的：更多开发自由软件的兴趣，更多自由软件商业的用户，以及更有利于鼓励公司开发商业自由软件而不是让软件产品私有化。

但是对软件的兴趣增长快于对其所基于的哲学的了解，这带来了麻烦。我们面对上边描述的挑战和威胁的能力依赖于坚决主张自由的意志。为了确定我们的社团拥有这个意志，我们需要在新的用户来到社团时向他们传播这样的思想。

但是我们没有做到这一点：吸引新用户加入社团的努力大大超越了教育他们成为我们社团的好公民的努力。我们需要做这两件事，而且我们也需要保持这两种努力的平衡。

"开放源码"

当 1998 年社团的一部分决定停止使用术语“自由软件”并改为说“开放源码软件”时，教导新用户有关自由的观念变得更加困难。

一些喜欢该术语的人意欲避免“自由”与“白送”的混淆——这是一个正当的目标。可是其他人，打算将激励了自由软件运动和 GNU 工程的原则精神抛到一边，反而迎合行政和商业用户，而这些用户中的许多人持有一种将利润置于自由、社团和原则之

principle. Thus, the rhetoric of "open source" focuses on the potential to make high quality, powerful software, but shuns the ideas of freedom, community, and principle.

上的意识形态。因而，"开放源码"的花言巧语集中在制作高质量、功能强大的软件的潜能上，但是避开自由、社团以及原则的思想。

The "Linux" magazines are a clear example of this—they are filled with advertisements for proprietary software that works with GNU/Linux. When the next Motif or Qt appears, will these magazines warn programmers to stay away from it, or will they run ads for it?

《Linux》杂志是这样一个清晰的例子——它们被与 GNU/Linux 合作运行的私有软件广告所充斥。当下一个 Motif 或 Qt 出现时，这些杂志将警告程序员们远离它还是为它们登载广告呢？

The support of business can contribute to the community in many ways; all else being equal, it is useful. But winning their support by speaking even less about freedom and principle can be disastrous; it makes the previous imbalance between outreach and civics education even worse.

商业支持可以以多种方式为社团做贡献；其他种类的支持也都一样，它是有益的。但是为了赢得他们的支持而少论及自由和原则可能损失惨重；它使得前述"超越和公民意识教育"之间的失衡变得愈加糟糕。

"Free software" and "open source" describe the same category of software, more or less, but say different things about the software, and about values. The GNU Project continues to use the term "free software", to express the idea that freedom, not just technology, is important.

"自由软件"和"开放源码"或多或少地描述了同一软件类别，但是它们谈论的是软件的不同方面和价值。GNU 工程继续使用"自由软件"这个术语来表达"不仅仅是技术，自由也是重要的"这一思想。

4.7 术语简介

1. CDR/RW (Compact Disc Read/Read Write)只读激光压缩/可读写盘
2. DVD (Digital Video Disc)数字视频光盘
3. USB (Universal Searial Bus)通用串行总线
4. RAR (Ration Archive) 档案压缩率
5. DOS (Disk Operation System)磁盘操作系统
6. ZIP (Zip Package)一种档案文件压缩格式
7. GUI (Graphic User Interface)图形用户接口
8. SFX (Self-extracting)自解压
9. AVI (Audio Video Interface)音视频接口（一种音视频文件格式）
10. MPEG (Moving Picture Experts Group)动态图像专家组标准
11. JPEG (Joint Photographic Experts Group)联合摄影（图像）专家组

Exercises（练习）

1. Match the explanations in column B with words and expressions in column A.（搭配每组中同意义的词或短语）

A	B	A	B
资源	File system	对象	CAD
操作系统	Office automation	资源管理	Control of I/O
作业管理	Resource	计算机辅助设计	Object
文件系统	Word processing	三维动画	Memory Management
办公自动化	Job Management	I/O 操作控制	GUI
文本处理	Internet	存储管理	Resource Management
互联网	Operating System	图形用户界面	3D animation

2. Choose the proper words to fill in the blanks.（选词填空）

(1) Operating system should meet certain specific standards as modern computer hardware's ________.

(2) Linux is a ________ system.

(3) Microsoft Office 2000 has five types of ________.

(4) A dynamic presentations include ________ features.

(5)________ is the graphic images created by a computer.

Words to be chosen from:（可选词）
sophistication computer graphics multitasking edition multimedia

3. Read the following sentences and write "T"for true and "F" for false.（对与错）

(1) When you purchase a new computer, an operating software suitable to the hardware must be bought.(　　)

(2) Office automation is the application of computer and communication technology.(　　)

(3) Today only a few of organizations use office automation hardware and software.(　　)

(4) Microsoft Excel 2000 is a spreadsheet program.(　　)

(5) One of the five functions of operating system is memory management.(　　)

(6) AutoCAD has the formidable two-dimensional function only, but three dimensional function.(　　)

4. Translate the following English sentences into Chinese.（将下面英语句子翻译为中文）

(1) The operating system provided one interactive contact surface for the computer hardware and the application procedure.

(2) Now the operating system for home computers is Windows.

(3) Windows 98 is an upgrade to Microsoft's Windows 95 operating system.

(4) Web designers use Flash to create navigation controls, animated logos, long-form

animations with synchronized sound, and even complete, sensory-rich Web sites .

(5) Image processing operations can be roughly divided into three major categories, Image Compression, Image Enhancement and Restoration, and Measurement Extraction.

5. Practical assignment: get familiar with computer operation system and latest application software.（实践作业：到当地电脑城了解计算机操作系统和最新应用软件。）

附文 4: Reading Material（阅读材料）

The Future of Computer Graphics

The computer revolution has been the fastest revolution in the history of humankind. In just a decade, the computer has become a part of our lives, at home, at work, at recreational centers: in fact, we just cannot seem to avoid them no matter what we do. Over the last few years, computer graphics has emerged in its own right as an area of research and production. It has evolved from crude line drawings to beautifully realistic images rivaling scenes we can see in nature. Today, the science of computer graphics is concerned with more than pretty pictures. While research continues to advance the state of art in fields such as image processing, 3D modeling, animation, and rendering, there is also an active thrust to bring computer graphics into our daily lives. We are indeed living in the computer graphics revolution age.

Where is computer graphics now, and where is it headed? Let us look at some of the key areas where CG is being used, and the areas where it is slowly gaining a stronger foothold.

The foremost area where CG is being used to its maximum is in entertainment. In the early 1980s, Lucas films started the trend with their trilogy Star Wars, for which computer scientists and artists skilled in 3D animation were used to generate much of the spectacular effects seen in these films. More recently in 1995, Pixar became the first production house to generate an entire movie using only computer graphics with Toy Story. Computer graphics is now used in almost every Hollywood movie, such as the dinosaurs of Jurassic Park, or more subtle composite effects as seen in Forest Gump. The effects seen in films are going to become more widespread and more realistic as research in the area continues.

Virtual reality (or VR for short) is kind of a buzzword these days in computer graphics. VR is artificial reality created by a computer that is so enveloping that it is perceived by the mind as being truly real. VR exists in many forms. A traditional view of virtual reality uses headsets and data gloves. The headset serves as the eyes and ears to your virtual world, projecting sights and sounds generated by the computer. The data glove becomes your hand, enabling you to interact with this simulated world. As you move your head around, the computer will track your motion and display the right image. VR is the most demanding application for computer graphics, requiring hardware and software capable of supporting real time 3D graphics. At the moment the most VR is being used to generate engulfing video games and interesting environments. The real interest of VR stems from the potential benefit to humanity. VR is being used to train pilots with the aid of flight simulators. VR will someday allow doctors to train on virtual human bodies, allow students to experiment with new ideas and concepts, and also let us explore strange new planets. People who cannot enjoy the real world due to physical handicaps can gain pleasure from enjoying simple pleasures like playing a golf game in a virtual world.

Hospitals and medical research labs are relying more and more on CG to display images of internal organs and tissues. As technology progress, it will soon be possible for doctors to view these organs as holographic images, as of the real thing were sitting in front of them. Medical imaging systems are all based on CG for their functioning. Research labs use CG to simulate molecular structures of DNA and other body building genes to explore the workings of our bodies and to experiment with them.

In this computer-simulated world, you are the artist with complete control over all the elements: the lights, models, and the camera. Based on your choice, the computer will generate the desired image for you.

General Features of Operating Systems

An operating system is a master control program which controls the functions of the computer system as a whole and the running of application programs. All computers do not use the same operating systems. It is therefore important to assess the operating systems used on a particular model before initial commitment because some software is only designed to run under the control of specific operating systems. Some operating systems are adopted as “industry standards” and these are the ones which should be evaluated because they normally have a good software base. The reason for this is that software houses are willing to expand resources on the development of application packages for machines functioning under the control of an operating system which is widely used. The cost of software is likely to be lower in such circumstances as the development costs are spread over a greater number of users, both actual and potential.

Mainframe computers usually process several application programs concurrently, switching from one to the other, for the purpose of increasing processing productivity. This is known as multiprogramming (multi-tasking in the context of microcomputers), which requires a powerful operating system incorporation work scheduling facilities to control the switching between programs.This entails reading in data for one program while the processor is performing computations on another and printing our results on net another.

In multi-user environment an operating system is required to control terminal operations on a shared access basis as only one user can access the system at any moment of time. Such systems also require a system for record locking and unlocking, to prevent one user attempting to read a record while another user is updating it, for instance. The first user is allocated control to write to a record (or file in some instances) and other users are denied access until the record is updated and unlocked.

Some environments operate in concurrent batch and real-time mode. This means that a “background” job deals with routine batch processing whilst the “foreground” job deals with real-time operations such as airline seat reservations, on-line looking of hotel accommodation, or control of warehouse stacks, etc. the real-time operation has priority, and the operation system interrupts batch processing operation to deal with real-time enquiries or file updates. The stage of batch processing attained at the time of the interrupt is temporarily transferred to backing storage. After the real-time operation has been dealt with, the interrupted program is transferred back to internal memory from backing storage, and processing recommences from a “restart” point. The operating system also copies to disk backing storage the state of the real-time system every few minutes to provide a means of “recovering” the system in the event of a malfunction.

An operation system is stored on disk and has to be booted into the internal memory where it must reside throughout processing so that commands are instantly available. The operating system commands may exceed the internal memory capacity of the computer in which case only that portion of the OS which is frequently used is retained internally, other modules being read in from disk as required. Many microcomputers function under the control of a disk operating system known as DOS.

Chapter 5 Programming Language（程序设计语言）

教学要求

本章面向编程爱好者，要求学生有一定的公共英语基础。

教学内容

选择介绍 C 语言、VB、Java2、SQL 等主流计算机语言的起源、特点、概念、关键字等。

教学提示

为学生学习语言课扫除障碍，做必要的知识准备。

5.1 Turbo C++3.0

C language is a system programming language for UNIX, developed by Dennes Ritchieas.

There are several types of integer with different size: floating point, pointer (indirection called in C language), arrays and structures, but there is no Booleans and sets, because C language is not strongly typed one. For example, some compilers do not insert run-time checks on array **subscript**, etc. Type conversion is permissive. Address arithmetic can be performed on pointers; null is **denoted** by a zero value.

C language has procedures and functions. Parameters are always passed by value. Thus, for a subprogram being operated on a given data structure, the pointer to this structure has to be passed.

C 语言是 Dennes Ritchieas 为 UNIX 而开发的一种系统编程语言。

在 C 语言中有几种不同大小的整数类型，有浮点型、指针（C 语言中叫作间接性）、数组和结构体，但没有布尔型和集合型，C 语言不是强类型的语言。例如，某些编译程序对数组下标并不插入运行时间的检查等，允许类型转换，地址运算可对指针执行；空类型用零指出。

C 语言有过程和函数，参数总是通过数值来传递。这样，对于在一个给定的数据结构上操作的子程序来说，指向该结构的指针就必须加以传递。

Turbo C++3.0 is just a tool for making **C language,** and we can type a program into it, get a program compiled and at last make it become an executable file.

Turbo C++3.0 仅仅是一个 C 语言的编写工具，我们可以用它来编写、编译程序并生成一个可执行的文件。

 **Key words:** subscript（下标），denote（表示），C language（C 语言）

 UNIX 是一种安全性能很高的网络操作系统。

5.1.1 Keyword（关键字）

Keyword is a character string, which has the specific significance in C language. Usually it also is called the **reserved** word. The definition **identifier** by programmer should not use them. The keywords of C language are divided into following several kinds:

- The storage class:

auto, external , register, static;

- The Data types class:

char, int, float, double, signed, unsigned , short, long, void, struct, union, typedef, enum, sizeof;

- The control statement class:

do, while, for, if...else, switch...case..., default, goto, continue, break, return.

在 C 语言中，关键字是指规定的具有特定意义的字符串，通常也称为保留字。用户定义的标识符不应与关键字相同。C 语言的关键字分为以下几类：

与存储类别有关的：

自动、外部、寄存器、静态；

与数据类型有关的：

字符型、整型、浮点型、双精度型、有符号型、无符号型、短整型、长整型、空类型、结构型、联合型、定义新类型、枚举型、长度；

与程序控制结构有关的：

直到型循环、当型循环、for 循环、选择语句、开关语句、缺省、转移、继续、中断、返回。

 **Key words:** reserve（预留），identifier（标识符）

 C 语言中一共有 32 个关键字，这些关键字是由 ANIS 推荐的。

5.1.2 The Data Types（数据类型）

C language introduced a concept of "*data types*" which are used to define a variable before

C 语言中，变量使用前必须做到先定义，后使用，在定义的时候又引入了"数

its use. The definition of a variable will assign storage for the variable and define the type of data that will be held in the location. So what data types are available? Here is a table:

据类型”的概念。变量在定义的时候，将被声明为某种类型和获取到存储空间，并保存在该空间中。因此，什么是变量类型呢？如下表所示。

	int（整数）	float（浮点型）	double（双精度浮点型）	char（字符型）
Function（功能）	used to define integer numbers（用于定义整数）	used to define floating point numbers（用于定义浮点数）	used to define BIG floating point numbers（用于定义高精度浮点数）	used to define a character（用于定义一个字符）
An example（例子）	Int a,b;	Float a;	Double a;	Char c1;

Notes

int 用于定义一个整型变量；float 用于定义一个单精度浮点型变量；double 用于定义一个双精度浮点型变量；char 用于定义一个字符变量。

5.1.3 The Debugging Surface of Turbo C3.0（Turbo C3.0 调试界面）

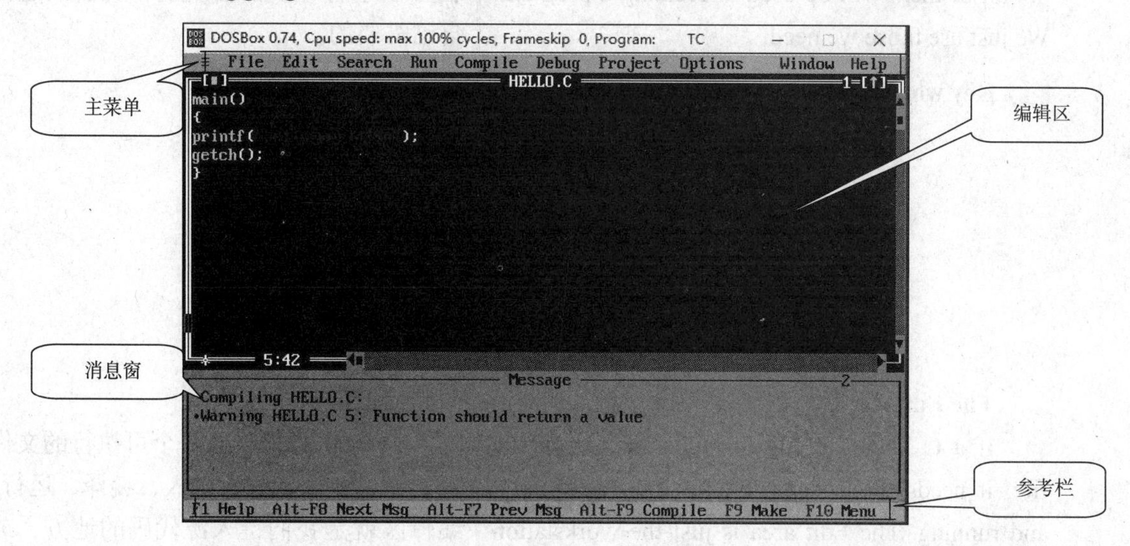

When the TC (Turbo C3.0 for short) is run, it will display the above interface. At the top of the windows is the menu; the middle of the window is the Edit area; the next one is the message area and the bottom of it is the reference column. These four parts form the main interface of Turbo C3.0.

进入 Turbo C3.0 集成开发环境中后，屏幕上会显示行如上面的图形界面。其中顶上一行为 Turbo C3.0 主菜单，中间窗口为编辑区，接下来是信息窗口，最底下一行为参考栏。这 4 个窗口构成了 Turbo C3.0 的主屏幕，之后的

Programming, compiling, **debugging** and running are all **executed** in it.

编程、编译、调试及运行都将在这个主屏幕中进行。

The Menu

At the top line of the main screen, it shows the following information: File, Edit, Search, Run, Compile, Debug, Project, Options, Window, and Help. They all have **submenu** except Edit. We can press “Alt” key plus the first letter of one of these items to access the submenu. For example, we can press “Alt+F” to choose the “File” submenu. All the submenus hold several relevant functions, but not all of them will be used in creating a program. We just use those we need.

主菜单

主菜单在主屏幕顶上一行，显示下列内容：File、Edit、Search、Run、Compile、Debug、Project、Options、Window 和 Help。除 Edit 外，其他各项均有子菜单，只要按 Alt 键加上某项中第一个字母，就可进入该项的子菜单中。例如，我们可以按下 Alt+F 组合键进入 File 子菜单。所有的子菜单都对应着一些功能，但并不是在编写程序的时候都会用上，我们只选择自己需要的。

 **Key words:** debug（调试），execute（执行），submenu（子菜单）

The Edit Area

If a C language file becomes an executable file, it needs three steps: typing into, compiling, and running. The Edit area is just the workstation where we type a raw code or where an existed file is opened. After finishing typing into, we should get it compiled successfully, if not, we have to debug it within the Edit area until all errors are corrected. So the Edit area is issue the information passage between the user and the system.

编辑区

一个 C 程序变成一个可执行的文件需要经过 3 个步骤：键入、编译、运行。编辑区就是我们键入源代码的地方，或者我们也可以在这里打开一个已经存在的文件。完成输入以后，我们就得调试程序直到成功，如果没有成功，我们必须在编辑区内更正所有的错误。因此，编辑区就是用户和系统信息交互的通道。

The Message Area

We have learned that some programs cannot be compiled successfully after typing into, and maybe they still have some errors. So we must debug it. But, how can we know which segment of code or word causes the errors? Now you should get the information from the message area, it will tell you where the errors occur, then correct these errors with the help of the message area's messages and get it compiled again until successfully.

The Reference Column

It is a grey bar at the bottom of the interface, it shows some shortcut keys.

消息区

我们已经知道，一些程序在键入后并不能顺利地执行，它们还有很多错误，因此，我们必须调试它。但是，我们怎么知道引起错误的是哪段代码或者哪个字呢？现在，你就应该读消息区的信息了，它会告诉你出错的具体地方，然后，你就借助这些提示信息，调试程序，直到成功为止。

参考栏

参考栏是位于界面下部的一个灰色长条，上面显示了一些快捷键。

5.2 Visual Basic 6.0

VB is Visual Basic for short, it is a kit empoldered by Microsoft corporation in the USA for **empoldering** softwares which runs under the Windows environment. It supplies a **visible** environment and the programming language is very easy, convenient to use. Here are some of its **excellences:**

1. platform for empoldering is visible;
2. **object-oriented** programming methods;

Visual Basic 简称 VB，是美国微软公司推出的 Windows 环境下的软件开发工具，它提供的是一种可视化编程环境，编程语言非常简单、使用方便，其优点有：

1. 可视化的编程平台；
2. 面向对象的设计方法；

3. **event-driven** programming methods;
4. using Windows's source sufficiently;
5. **structure** design language;
6. open database function and sustain network.

3．事件驱动的编程机制；
4．充分利用 Windows 资源；
5．结构化设计语言；
6．开放的数据库功能和网络支持。

Key words: empolder（开发），visible（可见的），excellence（优点），object-oriented（面向对象），event-driven（事件驱动），structure（结构化）

5.2.1 Some Common Events（常用事件）

Activate:

When one **object** becomes an active window, this **event occurs**.

Activate 事件:

当一个对象成为活动窗口时发生。

ButtonClick:

When the user **clicks** the object inside of the Toolbar control, this event occurs.

ButtonClick 事件:

当用户单击 Toolbar 控件内的按钮对象时发生。

Change:

When the contents of the some controls are changed by the user or the procedure code, this event occurs.

Change 事件:

当某个控件的内容被用户或程序代码改变时发生。

Click:

When the user clicks some object once with the mouse, this event occurs.

Click 事件:

当用户用鼠标单击某个对象时发生。

Dblclick:

When the user clicks some object twice with the mouse, this event occurs.

Dblclick 事件:

当用户用鼠标双击某个对象时发生。

Deactivate:

When one object isn't an active window, this event occurs.

Deactivate 事件:

当一个对象不再是活动窗口时发生。

DownClick:

When click arrow button which indicates down or left, this event occurs.

DownClick 事件:

单击向下或向左箭头按钮时发生。

ExitFocus:

When the focus leaves the object, this event occurs.

ExitFocus 事件:

当焦点离开对象时发生。

GotFocus:

When one object obtains the **focus**, this event occurs.

GotFocus 事件:

当对象获得焦点时发生。

Hide:

When the attribute value of the “visible” becomes “False”, this event occurs.

Hide 事件:

当对象的Visible属性变为False时发生。

ItemCheck:

When the Style **attribute** value of the **ListBox** control is established 1, and one check box of the ListBox control is chosen or is eliminated , this event occurs.

ItemCheck 事件:

当 ListBox 控件的 Style 属性设置为 1（复选框），并且 ListBox 控件中一个项目的复选框被选定或者被清除时发生。

KeyDown:

When one object has the focus and one key of the keyboard is pressed down, this event occurs.

KeyDown 事件:

当一个对象具有焦点时和按下键盘的一个键时发生。

KeyPress:

When one key of the keyboard is pressed down and then undone, this event occurs.

KeyPress 事件:

先按下再松开键盘上的一个键时发生。

KeyUp:

When one object has the focus, and a key is set free, this event occurs.

KeyUp 事件:

当一个对象具有焦点时和释放一个键时发生。

Load:

When one form is loaded, this event occurs.

Load 事件:

当窗体被装载时发生。

LostFocus:

When one object loses its focus, this event occurs.

LostFocus 事件:

当对象失去焦点时发生。

MouseDown:

When the user presses down the button of the mouse, this event occurs.

MouseDown 事件:

当用户按下鼠标按钮时发生。

MouseMove:

When the mouse is moving, this event occurs.

MouseMove 事件:

移动鼠标时发生。

MouseUp:

When the button of mouse is free, this event

MouseUp 事件:

释放鼠标按钮时发生。

occurs.

Paint:

After the movement or the enlargement of one object, or one from that covers the object is put aside and a part of or the entire object is shown, this event occurs. But, it doesn't occur when the attribute value of the AutoRedraw equals to "True".

PathChange:

When the attribute value of the "FileName" of "Path" is changed by the user, or the path of one file is changed, the event occurs.

PatternChange:

When the pattern of one file is changed, this event occurs.

Resize:

When the size of one object first displayed is changed, this event occurs automatically.

Scroll:

When the user drags the scroll bar with his mouse, this event occurs.

SelChange:

The choice of current text or the insert point is changed, this event occurs.

Show:

When the "Visible" attribute value of one object is changed into "True", this event occurs.

Timer:

After the value of the "interval" is past, this event occurs.

TimeChanged:

When the time is changed by the system or some applications, this event occurs.

Unload:

When one form is deleted from the screen, this event occurs.

Paint 事件：

在一个对象被移动或放大之后，或在一个覆盖该对象的窗体被移开之后，该对象部分或全部暴露时，此事件发生，但该事件在AutoRedraw属性设置为True时不发生。

PathChange 事件：

当用户指定新的 FileName 属性或Path属性，从而改变了路径时发生。

PatternChange 事件：

当文件的列表样式改变时发生。

Resize 事件：

当对象第一次显示的尺寸发生变化时自动发生。

Scroll 事件：

用户用鼠标在滚动条内拖动滚动框时发生。

SelChange 事件：

当前文本的选择发生改变或插入点发生变化时发生。

Show 事件：

当对象 Visible 属性变为 True 时发生。

Timer 事件：

计时器控件Interval预定时间过去后发生。

TimeChanged 事件：

应用程序或“控制面板”改变系统时间时发生。

Unload 事件：

当窗体从屏幕上删除时发生。

Key words: object（对象）, event（事件）, occur（发生）, click（鼠标点击）, focus（焦点）, attribute（属性）, ListBox（列表框，一种控件）

5.2.2 Some Common Controls，Attributes of Controls （常用控件及其属性）

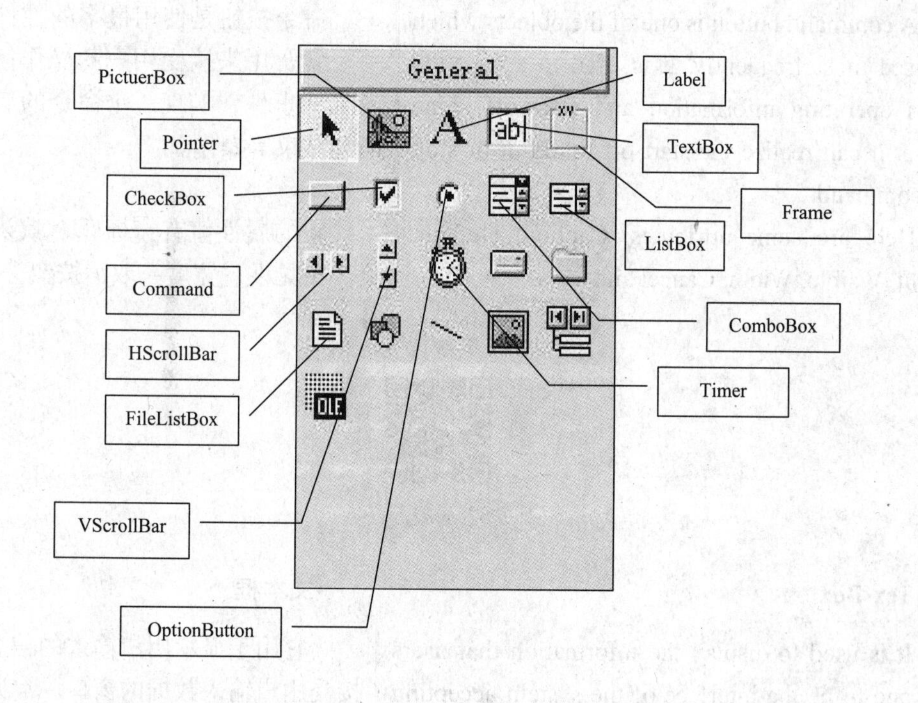

Form

Form is an object, which has its own attributes, functions, and events. It is a basic part of VB. User can get some information from the form; meanwhile, it is the carrier of some other objects. It owns the same characteristic like the windows. The attributes of form are: Caption, Height, Left, Top, Visible, Width, Backcolor, Enabled, FillColor, Font etc.

窗体

窗体是具有自身特定属性、功能和事件的一个对象。它是 VB 的一个基本构成部分，是运行程序时用户交互的实际窗口，也是其他对象的载体。它具有窗体的基本特性。常见的属性值有：标题、高度、左边距、上边距、可见性、宽度、背景色、有效性、填充色、字体等。

Notes

很多控件都具有相同的属性，如标题(caption)、高度(height)、宽度(width)、可见性(visible)、填充色 (fillcolor)。

Command

A command button is one of the objects which are used most frequently. It is used to accept the user's operating information and start off some events. It can realize the start off, intermit or stop of a command.

Here are some attributes: Caption, Default, Height, Visible, Width, Cancel and so on.

命令按钮

命令按钮是使用最多的控件对象之一。常常用来接受用户的操作信息，激发某些事件，实现一个命令的启动、中断和结束等操作。

它的属性值有：标题、默认值、高度、可见性、宽度、取消按钮等。

TextBox

It is used to display the information that users type and to be the interface of the system accepting the users' typed information. Here are some common attributes:

Text, alignMent, MaxLength, Locked, SelStart, SelLength, SelText, PasswordChar and so on.

文本框

它用于显示用户输入的信息，作为接受用户输入数据的窗口。常见的属性有：

文本、对齐、最多字数、编辑与否、选定文本长度、被选文本、密码字符等。

Label

It is used to display some text information, but it has no the function of input. It is mainly used to sign and display some hint information. Here are its common attributes:

Alignment, Caption, AutoSize, Name, Font, ForeColor, BackColor, Visible, Borderstyle etc.

标签

标签是用来显示文本的控件，但没有文本输入功能，主要用来标注和显示提示信息。其属性值有：

对齐方式、标题、自动调整大小、名称、字体、前景色、背景色、可见性、边框样式等。

ListBox and ComboBox

ListBox and ComboBox belong to the list controlling class. They are used to provide some choices for users. They have many similar functions, attributes, methods and events. Except for some basic attributes, they also have columns, lists, ListIndex and so on.

列表框和组合框

列表框和组合列表框都是列表类控件，向用户提供可选择项目的列表。它们有许多相似的功能、属性、方法和事件。常用的属性除了基本的属性以外还有：栏数、列表项、项目下标的索引等。

Option Button and CheckBox

There are all selective controls, and they have some differences and similarities. OptionButtons are often located in group. In some occasions, only one of buttons can be selected, but in CheckBox, you can select one, several, all buttons or none of them.

选择按钮和复选框

它们都是选择性的按钮，有相似的地方也有很多不同之处。选择按钮通常是以一组的形式出现的，有且只有一个能被选中，而复选框却可以同时选一个、几个、全部，或者一个也不选。

Option Button and CheckBox

Frame

It can be a container for some other controllers, and can also divide them into some marketable controller arrays. The objects in the frame should move with the frame's movement. But their locations are relative to the frame. Here are some attributes: caption, name, enable and so on.

框架

框架控件可以作为其他控件对象的容器，并将它们分成可标识的控件组。框架中的对象将随着框架移动，而其中对象的位置相对于框架也是不变的。常见的属性有名字、标题和是否可用。

Frame

Timer

It is an event controller which triggers off an event during a period of interval. Here are its two major attributes: enable and interval.

定时器控件

它是一种按一定时间间隔触发事件的控件，主要有两个属性：是否可用、间隔。

Timer

5.3 Java 8

The Java technology is one kind of computer programming tools, it was empoldered by a secret group that called "the green group" in Sun Microsystems Corporation in 1991. It was also a secret item named "the green plan". This secret "the green group" had 13 people leaded by James Gosling. They had locked themselves in an office without a name and shut off all normal contacts with Sun. They worked day and night for about 18 months.

Java 技术是一种计算机编程工具，是 Sun Microsystems 公司于 1991 年在一个名为"绿色团队"的小型秘密项目"绿色计划"中开发的。这个秘密的"绿色团队"共有 13 个人，由 James Gosling 领导。他们将自己锁在一个没有名字的办公室里，切断了与 Sun 公司的所有正常联系，然后夜以继日地工作了大约 18 个月。

In fact, the Java technology's multi-functions, validity, flexibility of working in different platforms as well as the security have already caused it to become the most perfect technology in network computation domain. Today, we can see Java which is widely applied everywhere: the Internet, the super science computers, the mobile phones, the family game machines and the credit cards.

事实上，Java 技术的多功能性、有效性、平台的可移植性以及安全性已经使它成为网络计算领域最完美的技术。今天，无论是互联网、科学超级计算机、手机，还是家庭游戏机和信用卡，在所有网络和设备上你都会看到 Java 技术的身影，它已经无处不在了。

Notes

印度尼西亚有一个重要的盛产咖啡的岛屿，中文名字叫爪哇，开发人员为这种新的语言起名为 Java，其寓意是为世人端上一杯热咖啡。Java 8 并不是新的语言，Java 8 的意思是 Java platform 8 version，是 Java 的不同版本罢了。

5.3.1 Java Language Keywords（Java 语言关键字）

The following is a list of keywords in the Java language. These words are reserved—you cannot use any of these words as names of variables in your programs. "True", "false", and "null" are not

下面有一张 Java 程序设计语言的关键字清单，这些字和 C 语言中的关键字一样，都是保留字，也就是说，你不能把它们作为变量名在你的程序中使用。

keywords, but they are also reserved words. So you cannot use them as names in your programs, either.

“True”“false” 和 “null” 虽然不是关键字，但是它们也是保留字，因此，在你的程序中你同样不能把它们作为名字。

Java Language Keywords1

Keywords:

abstract, continue, for, new, switch, assert, default, goto, package, synchronized, boolean, do, if, private, this, break, double, implements, protected, throw, byte, else, import, public, throws, case, enum, instanceof, return, transient, catch, extends, int, short, try, char, final, interface, static, void, class, finally, long, strictfp**, volatile, const, float, native, super, while

We have grasped 32 keywords in C language, so we can pick up some keywords from above very easily. For instance, they can be: continue, for, switch, default, goto, do and so on. But, do these words have same meanings and functions in C and Java? Yes. Some of them do. So, here we only learn some new keywords and the new functions of them in java.

在这之前我们已经在学习 C 语言的时候掌握了 32 个关键字，因此，我们可以很容易地从上面的关键字中找出在两种语言中都出现过的，例如，continue、for、switch、default、goto、do 等。但是，这些单词在两种语言中是否有相同的功能和用法呢？回答是肯定的，它们中有些单词具有相同的用法和功能。因此，我们在这儿就只介绍一些新的关键字和一些关键字的新功能了。

Java Language Keywords 2

Abstract:

This keyword is used in a class definition to specify that a class cannot be **instantiated**, but must be **inherited** by other classes. An **abstract** class can also have abstract methods that must be implemented by any concrete **subclasses**.

这个关键字用于类定义，指定某类不能作为具体的对象，而必须由其他类继承而来。抽象类的抽象方法可由具体子类执行。

Abstract

Byte:

A Java programming language keyword that represents a **primitive** data type that holds eight bits (binary digits) of data.

Java 程序语言的关键字，用于表示一个简单 8 位二进制的一种数据类型。

Byte

Catch:

A Java programming language keyword that declares a set of statements which will be executed if a matching Java **exception** occurs in the preceding “try” block.

Java 程序语言的关键字，用于定义一组语句，当前面的“try”块中，出现匹配异常时，就执行这些语句。

Catch

Implements:

A Java programming language keyword that must be included in the class declaration to indicate that the current classes are all implemented.

Java 程序语言的关键字，必须用在对类声明时，用于指明当前类都是接口成员。

Implements

Import：

It optionally appears before the class or interface declaration in the source file and is used to specify classes or packages which will be used or referred to later in the file without including their package names in the reference.

在源文件中出现在类或接口声明之前被用于指定类或包，这些包或者类在文件中以后要用到或者不在文件中却被提到。

Interface：

It is used to declare a class-like structure. An interface can declare attributes and methods, but it cannot provide method implementations. Any class implementing the interface is required to define the interface's method bodies.

它用于声明一个类似类的结构。这个接口只能声明属性和方法，但是它不能提供方法体。任何类执行这个接口，都必须定义这个接口的方法体。

New：

Java language keyword used to create an object from a class definition.

Java 程序语言的关键字用于创建一个类对象。

This:

A Java language keyword that **represents** the current class. It is used to access class variables and methods.

Java 程序语言关键字用于代表正在使用的当前类，用于访问类变量和类方法。

Throw:

It produces an exception to stop calling stack and **redirects** program execution to the first catch block.

它产生一个异常，阻止堆栈调用，并且将程序重定向到第一个 catch 块。

 **Key words:** instantiate（示例），inherit（继承，遗传得来的），abstract（抽象），subclass（子类），primitive（原始的，简单的），exception（异常），represent（代表），redirect（重定向）

5.3.2 Java Packages（Java 类库介绍）

Code **reuse** is an import concept in object-oriented programming. One tool that can help us increase the reusability of our code is the package statement. It is advised that you should split your entire codes, which is based on code functions, into specific application code and generic code. The application code which performs some special functions uses the generic code and finds the way to

代码重复使用在面向对象编程中是一个重要的概念。能够在我们的代码中帮助我们提高代码的可再用性的一种工具就是“包”。在编程的时候，经常被建议写程序时要基于代码功能将整个代码分为特殊应用代码和普通代码。完成特殊功能的程序代码一般支持普通代码，以便找到解决问题的捷径。

solve problems.

Most programming languages have a set of commands you can use to accomplish things. So does Java. But it is not the same as any others. Java's commands are all inside of classes that are contained in Java's packages.

多数编程语言有一套命令，你能用它们来完成一些事。Java 也有，但它和其他的语言有些不一样。Java 的命令全部包含在 Java 的“包”的类里面。

Do you still remember the first program of the Hello World when we start to learn Java first time? Compilers all begin with “import” which will tell the **compiler** that which parts of the Java's **API** should be loaded. For example, “import java.awt.Graphics;” means the compiler should load the graphics class.

你还记得我们第一次开始学习 Java 时的第一个程序“Hello World”吗？编译器都是从“import”开始的，并且他们会告诉编译器应该加载Java的API函数的哪部分。例如，“*import java.awt.Graphics;*”告诉编译器加载图形类。

The Packages

包

The classes of Java API are contained in the Java “Packages.” The packages are “java.lang”, “java.io”, “java.util”, “java.net”, “java.awt”, and “java.**applet**”. “Java.awt” has two sub-packages called “java.awt.peer” and “java.awt.image”.

Java API 函数的类，都包含在 Java 的包里。它们分别是“java.lang”“java.io”“java.util”“java.net”“java.awt”和“java.applet”。“java.awt”包含了两个子集，分别是“java.awt.peer”和“java.awt.image”。

“Java.lang” package is special because the classes contained in it are in your program automatically. They do not need an “import” statement.

“java.lang”是一个特别的类，因为这个包里的所有类都自动包含在你的程序中。它们不需要用到“import”语句。

java.awt.Graphics

You will use this class for drawing on your applet. The class has several methods for drawing .They are “drawString”, “drawRect”, and “fillRect”. All of these methods can match with relevant API reference. It is important to note that the Graphics class is abstract so you may not use the “new” operator with it. You may only turn variables into “Graphics” class with methods, and then the value to return is “Graphics” type.

在 applet 中作图时，将会用到这个类。这个类有几种作图的方法，分别是“drawString”“drawRect”和“fillRect”方法。所有这些方法都能找到相关的 API 函数。尤其值得注意的是，“Graphics”类是一个抽象类。因此，你不能将“new”和它一起使用。你只能将变量用一定的方法赋值成“Graphics”类型，这样返回的值才是“Graphics”类型。

朗读音频
Java.awt.Graphics

java.awt.Image

This class is obvious enough to hold a gif or jpeg picture. It contains several methods for **manipulating** the stored images. As for such class, it is very abstract, and an “Image” variable can only be **obtained** from a method and return with “Image type”.

这个类用于保存gif和jpeg格式图片显然足够了。它包含了几种方法用来对存储图片进行操作。像这样的类，十分抽象，并且图像变量只能从方法中获取并以“Image 类型”返回。

朗读音频
Java.awt.Image

java.lang.Math

This class contains important **mathematical** functions such as **square roots** and **logarithms**. You can use the class without defining a variable of “Math”.

这个类包含了十分重要的数学功能，如平方根和对数。你不需要定义一个“Math”类的变量来使用这个类。

java.net.URL

This class contains a Universal Resource Locator (URL), or the Location of a Document on the Internet. For example, URL should be the home page of the maker of http://Java.Sun.com, and the URL of this page is probably put in a text box at the top of your browser window.

这个类是包含在“统一资源定位”里，或者说成“文档在因特网中的位置”这个包里。举个例子，URL 应该是 http://Java.Sun.com 开发者的主页。这个页面的URL应该写在浏览器最上面的这个文档框里。

Key words: reuse（重复），compiler（编译器），API（Application Programming Interface 的缩写形式，应用编程接口），applet（Java 的一种程序），manipulate（操作），obtain（获得），mathematical（数学的），square root（平方根），logarithm（对数）

5.4 SQL—Data Concentration Camp（数据集中营）

SQL (Structure Query Language) was put forward by Boyce and Chamberlin in 1974. From 1975 to 1979, San Jose Research Laboratory of IBM developed the famous management system prototype of relation database — System R and realized it. Because its function is very powerful and the language is simple and direct, it is very popular among computer users and computer industrial circles. So it is adopted widely by numerous computer companies and software companies.

SQL (Structure Query Language)语言是1974年由Boyce和Chamberlin提出的。1975年至1979年间，IBM公司San Jose Research Laboratory研制了著名的关系数据库管理系统原型——System R 并实现了这种语言。由于它功能丰富，语言简洁，备受用户及计算机工业界欢迎，被众多计算机公司和软件公司所采用。

With the constant modification, expanding and improvement by every company, SQL language has developed into the standard language of the relation **database** finally.

经各公司的不断修改、扩充和完善，SQL 语言最终发展成为关系数据库的标准语言。

In October of 1986, X3H2 Database Committee of American National Standard Institute (abbreviated as ANSI) approved SQL as the American standard of relation database. In the same year, the standard text of SQL (abbreviated as SQL-86) was announced. In 1987, International Organization for Standardizations (abbreviated as IOS) agreed to this standard, too. With the constant revision and improvement, a series of standards，such as SQL-89, SQL-92 were announced by IOS and they influenced the fields beyond the database, too. The inquiring function of SQL was connected with figure functions, software project tools, software developing instruments, artificial intelligence procedure in many software products. SQL has already become a major language in database field.

1986 年 10 月，美国国家标准局（American National Standard Institute，简称 ANSI）的数据库委员会 X3H2 批准了 SQL 作为关系数据库语言的美国标准，同年公布了 SQL 标准文本（简称 SQL-86）。1987 年，国际标准化组织（Interna-tional Organization for Standardization，简称 IOS）也通过了这一标准，此后 ANSI 不断修改和完善 SQL 标准，并于 1989 年公布了 SQL-89 标准，1992 年又公布了 SQL-92 标准。SQL 成为国际标准，对数据库以外的领域也产生了很大的影响，有不少软件产品将 SQL 语言的数据查询功能与图形功能、软件工程工具、软件开发工具、人工智能程序结合起来。SQL 已成为数据库领域中一种主流语言。

 **Key words:** SQL (Structure Query Language)，database（数据库）

朗读音频
SQL-Data Concentration Camp

5.5 Python

Python is an **interpreted** computer programming language which was invented by Guido van Rossum in 1989, and it was first published in 1991. The Python **source** code and interpreter follow the GPL (GNU General Public License) protocol. The Python

Python 是一种解释型计算机程序设计语言，由 Guido van Rossum 于 1989 年发明，并于 1991 年首次发布。Python 源代码和解释器遵循 GPL (GNU General Public License)协议。

interpreter is available for many operating systems, and it allows Python code to run on various systems.

Python解释器可用于许多操作系统，允许Python代码在各种系统上运行。

Python **grammar** is concise and clear, one of the characteristics is to force the whitespace to be used as a statement indent.

Python语法简洁清晰，特色之一是强制用空白符作为语句缩进。

Python has a abundant and powerful library.It is often called glue language which can easily connect the various modules (especially C / C ++) which were made in other languages. A common application status is that Python is used to quickly build prototypes of the program, and then the special requirement parts will be rewrited in a more appropriate language. Such as the graphics **rendering** module in the 3D game, which has high performance requirements, can be rewrited in C/C++ languages.

Python具有丰富和强大的库。它常被称为胶水语言，能够把用其他语言制作的各种模块（尤其是C/C++）很轻松地联结在一起。常见的一种应用情形是，使用Python快速生成程序的原型，然后对其中有特别要求的部分，用更合适的语言改写，比如3D游戏中的图形渲染模块，性能要求特别高，就可以用C/C++重写。

Because of its open source nature, Python has been ported to many platforms. These platforms include Linux, Windows, FreeBSD, Macintosh, Solaris, OS / 2, Amiga, AROS, AS / 400, BeOS, OS / 390, z / OS, Palm OS, QNX, VMS, Psion, Acom RISC OS, VxWorks, PlayStation, Sharp Zaurus, Windows CE, PocketPC, Symbian, as well as android platform which was developed by Google.

由于它的开源本质，Python已经被移植在许多平台上。这些平台包括Linux、Windows、FreeBSD、Macintosh、Solaris、OS/2、Amiga、AROS、AS/400、BeOS、OS/390、z/OS、Palm OS、QNX、VMS、Psion、Acom RISC OS、VxWorks、PlayStation、Sharp Zaurus、Windows CE、PocketPC、Symbian以及Google开发的android平台。

Notes

be available for...: 对……有效

Key words: interpreted（解释）, source（来源，根源）, grammar（语法）, rendering（渲染）

5.6 Situation Dialogue（情境对话）

Clumsy Panda's Adventure in Programming Kingdom

笨熊猫程序园历险记

Today Tu Tu and Clumsy Panda have visited programming kingdom. Clumsy Panda is very happy because he has known many members in our programming language family. Look! They are

今天，图图和熊猫笨笨游览了程序乐园，笨笨可高兴了，因为它认识了我们程序设计语言大家族中的许多成员，你看，它们过来了。

coming.

Big elder brother—Machine language

Instructing system is also called machine language. Each instruction replies a string of binary system code. Machine language is the only language that a computer can identify and carry out directly. Compared with other programming languages, it is carried out with higher efficiency. Since machine language instruction is a string of binary system code, it lacks readability and cannot be easily memorized. Further more, it is very difficult to compile programs and the modification and debug of the programs also very hard. So, machine language is not easy to grasp and use.

大哥——机器语言

指令系统也称机器语言，每条指令都对应一串二进制代码。机器语言是计算机唯一能识别并直接执行的语言，所以与其他程序设计语言相比，其执行效率高。由于机器语言指令都是一串二进制代码，可读性差、不易记忆；编写程序既难又繁，容易出错；程序的调试和修改难度也很大，所以机器语言不容易掌握和使用。

Younger brother—Assembly language

Assembly language appeared at the beginning of 1950s. It is also called Symbol Language because it consists of some symbols which are very easy to identify and memorize. The program compiled with assembly language is called source program of assembly language. This kind of language can not be identified and carried out directly by computers and it must be firstly translated into machine language program (named object program). Only after that, it can be carried out then.

二弟——汇编语言

20 世纪 50 年代初出现了汇编语言，它使用的是比较容易识别、记忆的助记符号，所以汇编语言也叫符号语言。用汇编语言编写的程序称为汇编语言源程序，计算机不能直接识别和执行它，必须先把汇编语言源程序翻译成机器语言程序（称目标程序），然后才能被执行。

Little sister—High-grade programming language

Up to the middle of 1950s, people created high-grade programming language. High-grade language is a kind of language that is used to compile various programs with meaningful “words” and “mathematics formulas” according to certain “grammar rules”, also called high-grade programming language or algorithm language, including C, C++, Visual C and Visual Basic etc. Now, popular high-grade language is to adopt the method of compiling. Simply speaking, a high-grade language program can become executable machine language program only after being compiled and assembled.

小妹——高级程序设计语言

到了 20 世纪 50 年代中期，人们又创造了高级程序设计语言。高级语言是一种用各种意义的“词”和“数学公式”按照一定的“语法规则”编写程序的语言，也称高级程序设计语言或算法语言。目前流行的高级语言，如 C、C++、Visual C++、Visual Basic 等都是采用编译的方法。简单地说，一个高级语言源程序必须经过编译和连接装配两步后才能成为可执行的机器语言程序。

Clumsy Panda is awfully happy because he have gained so much today.

今天的收获太大了，笨笨心里高兴极了。

5.7 Reading and Compacting（对照阅读）

Object-Oriented Programming Languages

Object-oriented programming (OOP) languages like C++ are based on traditional high-level languages, but they enable a programmer to think in terms of collections of cooperating objects instead of lists of commands. Objects, such as a circle, have properties such as the radius of the circle and the command that draw it on the computer screen. Classes of objects can inherit features from other classes of objects. For example, a class defining squares can inherit features such as right angles from a class defining rectangles. This set of programming classes simplifies the programmer's task, resulting in more reliable and efficient programs.

SIMULA was the first object-oriented programming language. It was developed in the mid to late 1960s in Norway. Smalltalk, the language that popularized object-oriented concepts, was developed in the early 1970s.

The artificial intelligence research community embraced this new programming technology early on: many flavors and dialects of the LISP programming language provide object-oriented extension. In the 1970s, these languages were available only within research laboratories. At the beginning of the 1980s came the real dawn of the object-oriented programming

面向对象程序语言

像 C++这样的面向对象的程序设计语言（OOP）是以传统的高级语言为基础的，但是它们能使程序员按照组合对象集方式而不是指令列表方式来进行思考。对象有许多性质，以圆为例，就有圆的半径以及把圆画到计算机屏幕上的命令。对象的类可以从其他对象类那里继承属性。例如，一个定义正方形的类能从定义长方形的类那里继承诸如直角这样的属性。程序类的这种关系简化了程序员的工作，从而导致更多既可靠又高效的程序产生。

SIMULA 是第一个面向对象的程序设计语言，它是 20 世纪 60 年代中后期在挪威开发的。Smalltalk 是 70 年代初期开发的，它使得面向对象的概念大众化了。

这一新的程序设计技术很早便受到人工智能研究协会的欢迎：LISP 程序设计语言中的许多风格及惯用语提供了面向对象的扩展版本。20 世纪 70 年代这些语言只能在研究室内使用，80年代初期真正开始了面向对象的程序设计时代。

languages.

In 1983, Smalltalk 80 realized commercialization, other object oriented programming languages, such as Objective-C, Eiffel, the Common Lisp Object System, and Actor, which have become commercially available.

1983年，Smalltalk 80实现了商品化，其他面向对象的程序设计语言，如Objective-C、Eiffel、Common Lisp Object System 和Actor也已商品化。

The long-term productivity of systems is enhanced by object-oriented program. Because of the modular nature of the code, programs are more malleable. This is particularly beneficial for applications that will be used for many years, during which companies' needs may change and making software modifications become necessary. Software reliability can be improved by object-oriented programming. Since the objects are repeatedly tested in a variety of applications, bugs are more likely to be found and corrected. Object-oriented programming also has potential benefits in parallel processing. Execution speed of parallel processing under object oriented methods will be improve.

面向对象的程序设计提高了系统的长期生产率，这种程序模块的本质是使程序更易适应。这一点对那些需要使用很多年的应用程序来说尤其有益，因为在这期间公司的需求可能有变化，软件的修改是必不可少的。面向对象的编程能提高软件的可靠性。由于对象在不同的应用中受到反复检验，更有可能将错误查出并纠正。面向对象的编程在并行处理中也有潜在优势，使用面向对象的方法使并行处理的执行速度将得到提高。

5.8 术语简介

1. IBM：International Business Machine Corp（国际商用机器公司）。

2. BASIC：Beginners Allpurpose Symbolic Instruction Code（初学者通用符号指令码）。

3. UNIX：Uniplexed Information and Computer Systems, UNIX操作系统，1969年在AT&T Bell实验室开发的多用户多任务操作系统。

4. C：C语言，一种高级程序设计语言，由贝尔实验室开发成功。

5. DBMS: Database Management System （数据库管理系统）。

6. Pascal: Philips Automatic Sequence Calculator（菲利浦自动顺序计算机语言）。

7. Operand: 操作数；运算数。

8. COBOL: Common Business-Oriented Language（面向商业的通用语言），COBOL语言。

Exercises（练习）

1. Translate the following phrases into English.（将下面短语翻译为英文）

（1）机器码

（2）机器语言

（3）面向对象的程序

（4）汇编语言
（5）抽象代码
（6）人工智能
（7）并行进程
（8）数据库系统

2. Please list the computer advanced programming languages that are used frequently and make some relevant introductions.（请列出一些经常使用的计算机高级编程语言，并做相关介绍）

附文5: Reading Material（阅读材料）

Today's Database Landscape

Computing technology has made a permanent change in the ways businesses work around the world. Information that was at one time stored in warehouses full of filing cabinets can now be accessed instantaneously at the lick of a mouse button. Orders placed by customers in foreign countries can now be instantly processed on the floor of a manufacturing facility. Although 20 years ago much of this information had been transported onto corporate mainframe databases, offices still operated in a batch-processing environment. If a query needed to be performed, someone notified the management information systems (MIS) department; the requested data was delivered as soon as possible (though often not soon enough).

In addition to the development of the relational database model, two technologies led to the rapid growth of what are now called client/server database systems. The first important technology was the personal computer. Inexpensive, easy-to-use applications such as Lotus 1-2-3 and Word Perfect enabled employees (and home computer users) to create documents and manage data quickly and accurately. Users became accustomed to continually upgrading systems because the rate of change was so rapid, even as the price of the more advanced systems continued to fall.

The second important technology was the local area network (LAN) and its integration into offices across the world. Although users were accustomed to terminal connections to a corporate mainframe, now word processing files could be stored locally within an office and accessed from any computer attached to the network. After the Apple Macintosh introduced a friendly graphical user interface, computers were not only inexpensive and powerful but also easy to use, in addition, they could be accessed from remote sites, and large amounts of data could be off-loaded to departmental data servers, during this time of rapid change and advancement, a new type of system appeared, called client/server development because processing is split between client computers and a database server, this new breed of application was a radical change from mainframe-based application programming. Among the many advantages of this type of architecture are:

- Reduced maintenance cost.
- Reduced network load (processing occurs on database server or client computer).
- Multiple operating systems that can interoperate as long as they share a common network protocol.
- Improved data integrity owing to centralized data location.

In Implementing Client/Server Computing, Bernard H.Boar defines client/server computing as follows:

Client/Sever computing is a processing model in which a single application is partitioned between multiple processors (front-end and back-end) and the processors cooperate (transparent to the end user) to complete the processing as a single unified task. Implementing Client/Server Computing A client/server bond product ties the processors together to provide a single system image (illusion). Shareable resources are positioned as requestor clients that access authorized services. The architecture is endlessly recursive; in turn, servers can become clients and request services of other servers on the network, and so on.

This type of application development requires an entirely new set of programming skills. User interface programming is now written for graphical user interfaces, whether it is MS Windows, IBM OS/2, Apple Macintosh, or the UNIX X-Window system. Using SQL and a network connection, the application can interface to a database residing on a remote server. The increased power of personal computer hardware enables critical database information to be stored on a relatively inexpensive standalone server. In addition, this server can be replaced later with little or no change to the client applications.

Chapter 6 Computer Network Technology（计算机网络技术）

教学要求

掌握专业关键词汇（key words）；能阅读本章所列英语短文；了解网络组件和应用。

教学内容

网络组件；各相关网络设备品牌；网络应用；常用的专业术语。

教学提示

到学校机房或本地电脑城参观，感受本章内容，以学到更多的专业词汇。

ZTE中兴

6.1 5–Layer TCP/IP Model（TCP/IP 模型）

The basic structure of communication networks is represented by the **Transmission Control Protocol/Internet Protocol (TCP/IP)** model. This model is structured in five layers. An end system, an intermediate network node, or each communicating user or program is equipped with devices to run all or some portions of these layers, depending on where the system operates. These five layers are as shown in the figure.

通信网络的基本结构是以传输控制协议/互联网协议（TCP/IP）模型为代表的，该模型是 5 层结构。一个终端系统、一个网间节点，或者每个通信用户或程序都配备有运行这些层的所有或者某些部分的设备，这取决于该系统在哪里运行。5 层结构如下页图所示。

Layer 1, the physical layer, defines electrical aspects of activating and maintaining physical links in networks. The physical layer represents the basic network hardware, such as switches and routers.

第 1 层，物理层，对激活和维护网络物理链路的电器方面进行了定义。这一层代表了物理层基本的网络硬件，如交换机和路由器。

Layer 2, the link layer, provides a reliable synchro- nization and transfer of information across the physical layer for accessing the transmission medium. Layer 2 specifies how packets access links and are attached to additional headers to form frames when entering a new networking environment, such as a LAN. Layer 2 also provides error detection and flow control.

第 2 层，链路层，为访问传输介质提供可靠的同步和物理层的信息传输。第二层指定数据包如何访问链路，并附加额外的头信息，这样当进入新的网络环境，如 LAN，就形成了帧。第 2 层还提供差错检测和流量控制。

Layer 3, the network layer (IP), specifies the networking aspects. This layer handles the way that addresses are assigned to packets and the way that packets are supposed to be forwarded from one **end** point to another.

第 3 层，网络层（IP），用于指定网络方面的功能。这一层处理的方法是把地址分配给数据包，并将数据包由一个终端转发到另一个终端。

Layer 4, the transport layer, lies just above the network layer and handles the details of data transmission. Layer 4 is implemented in the end-points but not in network routers and acts as an interface protocol between the communication terminal and network.

第 4 层，传输层，位于上述网络层之上并且处理数据传输的细节。第 4 层是在终端执行而不是在网络路由器上，并且作为一个接口协议作用在通信终端和网络之间。

Layer 5, the application layer, determines how a specific user application should use a network. Among such applications are the **Simple Mail**

第 5 层，应用层，决定了一个特定的用户应用程序应如何使用一个网络。这些应用是简单邮件传输协议

Transfer Protocol (SMTP), **File Transfer Protocol (FTP)**, and the **World Wide Web (WWW)**.

SMTP、文件传输协议 FTP 和万维网 WWW。

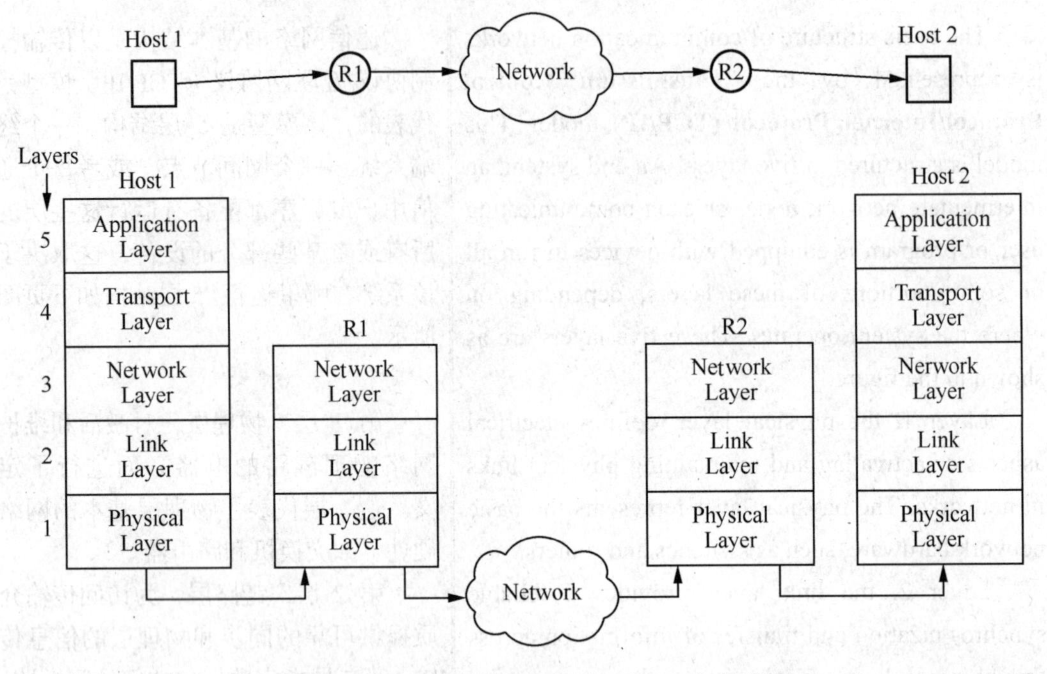

Hierarchy of 5-layer communication protocol model

The transmission of a given message between two users is carried out by (1) flowing down the data through each and all layers of the transmitting end, (2) sending it to certain layers of protocols in the devices between two end points, and (3) when the message arrives at the other end, letting the data flow up through the layers of the receiving end until it reaches its destination. The figure illustrates a scenario in which different layers of protocols are used to establish a connection. A message is transmitted from host 1 to host 2, and, as shown, all five layers of the protocol model participate in making this connection. The data being transmitted from host 1 is passed down through all five layers to reach router R1.

两个用户之间的给定信息的传输是通过以下步骤实现的：（1）发送端的数据依次通过每一层传输下来；（2）在两个端点的设备之间发送传输下来的数据至协议的特定层；（3）当消息到达另一端时，让通过接收端的数据流（或传输下来的数据）通过接收端的那些层直到传输至目的地。上图说明了在不同层的协议用于建立连接的情况。从主机 1 发送一条消息至主机 2，协议模型的 5 层都参与建立这一连接。传输的数据从主机 1 经过所有的 5 层，到达了路由器 R1。

Router R1 is located as a gateway to the operating regions of host 1 and therefore does not involve any tasks in layers 4 and 5. The same scenario is applied at the other end: router R2. Similarly, router R2, acting as a gateway to the operating regions of

路由器 R1 所处的位置就像通往主机 1 工作区域的大门，因此并不涉及在第 4 层和第 5 层的任何任务。同样的情形也适用于另一端：路由器 R2。路由器 R2 在通往主机 2 的工作

host 2, does not involve any tasks in layers 4 and 5. Finally at host 2, the data is transmitted upward from the physical layer to the application layer.

区域也起到了桥梁的作用，也并不涉及在第 4 层和第 5 层的任何任务。最后，主机 2 从物理层到应用层向上传输数据。

 **Key words:** TCP/IP（Transmission Control Protocol/Internet Protocol，传输控制协议/互联网协议），end（终端），SMTP（Simple Mail Transfer Protocol，简单邮件传输协议），FTP（File Transfer Protocol，文件传输协议），WWW（World Wide Web，万维网）

6.2 7-Layer OSI Model（OSI 模型）

The **open systems interconnection (OSI)** model was the original standard description for how messages should be transmitted between any two points. To the five TCP/IP layers, OSI adds the following two layers:

Layer 6, the session layer, which sets up and coordinates the applications at each end.

Layer 7, the presentation layer, which is the operating system part that converts incoming and outgoing data from one presentation format to another.

The tasks of these two additional layers are dissolved into the application and transport layers in the newer five-layer model. The OSI model is becoming less popular. TCP/IP is gaining more attention, owing to its stability and its ability to offer a better communication performance.

OSI and TCP/IP correspondence shown in the figure.

开放系统互连（OSI）模型是最初的标准，它描述的是在任意两点之间应如何传输信息。相对于 5 层 TCP / IP 模型，OSI 模型添加了以下两层：

第 6 层，会话层，安排和协调两端的应用。

第 7 层，表示层，它是操作系统的一部分，把输入和输出数据从一种表示格式转换到另一种格式。

这两个额外的层的任务被渐渐划入 5 层模型的应用层和传输层。OSI 模型正越来越不受欢迎，TCP/IP 则获得了更多的关注，这是由于它具有很好的稳定性，还能提供更好的通信性能。

OSI 模型和 TCP/IP 模型的对比见下图。

 **Key words:** open systems interconnection（OSI，开放系统互连）

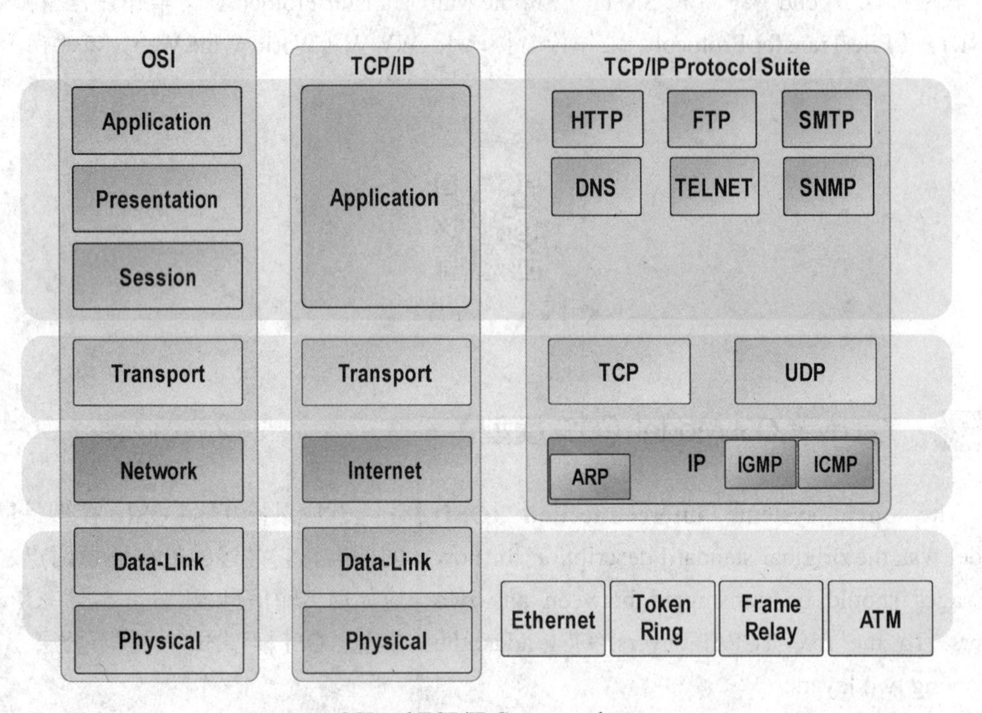

OSI and TCP/IP Correspondence

6.3 Networks Components（网络组件）

6.3.1 Switch（交换机）

A network **switch** is a device that **forwards** and **filters** OSI layer 2 **datagrams** (chunk of data communication) between **ports** (connected cables) based on the MAC addresses in the packets. This is distinct from a hub in that it only forwards the **frames** to the ports involved in the communication rather than all ports connected. A switch breaks the collision domain but represents itself a **broadcast domain**. Switches make forwarding decisions of frames on the basis of MAC addresses. A switch normally has numerous ports, facilitating a star topology for devices. Some switches are

网络交换机是一种依据数据包的 MAC 地址在端口（连接电缆）转发和过滤 OSI 第二层数据包（数据通信块）的设备。交换机只转发帧至涉及的端口，而不是所有的连接端口，这是不同于集线器的。交换机隔离了冲突域，但它本身就是一个广播域。交换机依据 MAC 地址决定转发帧。交换机通常有多个端口，这样比较适合用星型拓扑。有些交换机具有基于第 3 层寻址或

capable of routing based on Layer 3 addressing or additional logical levels; these are called multi-layer switches. The term switch is used loosely in marketing to encompass devices including routers and bridges, as well as devices that may distribute traffic on load or by application content (e.g., a Web URL identifier).

其他逻辑层的路由能力，这些被称为多层交换机。交换机这个术语宽泛地用在营销中，包括路由器和网桥设备，以及可以分配负载流量或应用内容（例如，一个 WebURL 标识符）的设备。

Key words: switch（交换机），forward（转发），filter（过滤），datagram（数据包），port（端口），frame（帧），broadcast domain（广播域）

常见的交换机厂商：思科（Cisco）、Juniper、神州数码（Digital China）、锐捷网络、华为（HUAWEI）、友讯网络（D-Link）、中兴（ZTE）。

Switch

Router

朗读音频

Switch

6.3.2 Router（路由器）

A **router** is a networking device that forwards packets between networks using information in protocol headers and **forwarding tables** to determine the best next router for each packet. Routers work at the Network Layer (layer 3) of the OSI model and the Internet Layer of TCP/IP.

路由器是一种用于转发网络间数据包的设备，通过使用协议头和转发表的信息，以确定最适合每个包的下一个路由器。路由器工作在 OSI 模型的网络层（第 3 层）和 TCP/IP 的网络层。

Key words: router（路由器），forwarding table（转发表）

常见的路由器厂商：思科（Cisco）、Juniper、神州数码（Digital China）、锐捷网络、华为（HUAWEI）、友讯网络（D-Link）、中兴（ZTE）

Router

6.4 Applications of Internet（Internet 应用）

6.4.1 HTTP（超文本传输协议）

Until the 1990s the Internet was used primarily by researchers, academics, and university students to log-in to remote hosts, to transfer files from local hosts to remote hosts and vice versa, to receive and send news, and to receive and send electronic mail. Although these applications were (and continue to be) extremely useful, the Internet was essentially unknown outside the academic and research communities. Then, in the early 1990s, the Internet's killer application arrived on the scene—the World Wide Web. The Web is the Internet application that caught the general public's eye. It is dramatically changing how people interact inside and outside their work environments. It has spawned thousands of start up companies.

20 世纪 90 年代以前，互联网主要是科研人员和大学师生用于登录到远程主机，在本地主机和远程主机间传送文件、收发新闻和电子邮件的。尽管这些应用在当时极为有用，但互联网还基本上不为学术和科研群体以外的人所知。后来，到 20 世纪 90 年代早期，互联网的重量级应用 Web 面世。Web 是一个普及型的互联网应用。它深刻地改变了人们在工作环境内外的交互手段，造就了数千个新兴的公司。

History is sprinkled with the arrival of electronic communication technologies that have had major societal impacts. The first such technology was the telephone, invented in the 1870s. The next electronic communication technology was broadcast radio/television, which arrived in the 1920s and 1930s. The third major communication technology that has changed the way people live and work is the Web. Perhaps what appeals the most to users about the Web is that it operates **on demand**. Users receive what they want, when they want it. This is unlike broadcast radio and television. In addition to being on demand, the Web has many other wonderful features that people love and cherish. It is enormously easy for any individual to make any information available over the Web. **Hyperlinks** and **search engines** help us navigate through an ocean of **Web sites**. Graphics and animated graphics stimulate our senses. Forms, Java applets, Active X components, as well as many other devices enable us to interact with pages and sites. And more and more, the

历史上，先后问世了多个具有重大社会影响的电子通信技术。第一个这样的技术是 19 世纪 70 年代发明的电话。下一个电子通信技术是 20 世纪 20 年代及 30 年代问世的广播收音机/电视机。改变人们的生活与工作方式的第三个重大通信技术就是 Web。Web 最吸引用户的也许是它的随选操作性，用户只在想要时收到所要的东西，这一点不同于广播收音机和电视机。除了随选操作性，Web 还有许多大家喜爱的其他精彩特性：任何个人都可以极其容易地在 Web 上公布任何信息；超链接和搜索引擎帮助我们在 Web 站点的海洋中导航；图形和动画刺激着我们的感官；表单、Java 小应用程序、Active X 控件以及其他许多设备使我们能与

Web provides a menu interface to vast quantities of audio and video material stored in the Internet, audio and video that can be accessed on demand.

Web页面和站点交互；Web还越来越普遍地提供存放在互联网中的、可随选访问（即点播）的大量音频和视频材料的菜单接口。

The **Hypertext Transfer Protocol (HTTP)**, the Web's application-layer protocol, is at the heart of the Web. HTTP is implemented in two programs: a client program and a server program. The client program and server program, executing on different end systems, talk to each other by exchanging HTTP messages. HTTP defines the structure of these messages and how the client and server exchange the messages.

Web的应用层协议HTTP（超文本传输协议）是Web的核心。HTTP在Web的客户端程序和服务器程序中得以实现，运行在不同终端系统上的客户端程序和服务器程序通过交换HTTP消息彼此交流，HTTP定义这些消息的结构以及客户和服务器如何交换这些消息。

 **Key words:** frame relay（帧中继），on demand（随选），hyperlink（超链接），search engine（搜索引擎），Web site（Web站点），HTTP（Hypertext Transfer Protocol，超文本传输协议）

6.4.2 Electronic Mail（电子邮件）

Along with the Web, **electronic mail** is one of the most popular Internet applications. Electronic mail is fast, easy to distribute, and inexpensive. Moreover, modern electronic mail messages can include hyperlinks, HTML formatted text, images, sound, and even video. In this section we will examine the application-layer protocols that are at the heart of Internet electronic mail.

电子邮件是互联网上最为流行的应用之一。电子邮件既迅速，又易于发送，而且成本低廉。另外，现代的电子邮件消息可以包含超链接、HTML格式文本、图像、声音，甚至视频数据。我们将在本节学习处于互联网电子邮件核心地位的应用层协议。

The **Simple Mail Transfer Protocol (SMTP)** is the principal application-layer protocol for Internet electronic mail. It uses the reliable data transfer service of TCP to transfer mail from the sender's mail server to the recipient's mail server. As with most application-layer protocols, SMTP has two sides: a client side, which executes on the sender's mail server, and a server side, which

简单邮件传输协议（SMTP）是互联网电子邮件系统重要的应用层协议，它使用由TCP提供的可靠的数据传输服务把邮件消息从发信人的邮件服务器传送到收信人的邮件服务器。跟大多数应用层协议一样，SMTP也存在两个端：在发信人的邮件服务器上执行的客户端和在收信人的邮件服务器上执行的服务器

executes on the recipient's mail server. Both the client and server sides of SMTP run on every mail server. When a mail server sends mail (to other mail servers), it acts as an SMTP client. When a mail server receives mail (from other mail servers) it acts as an SMTP server.

端。SMTP的客户端和服务器端同时运行在每个邮件服务器上。当一个邮件服务器在向其他邮件服务器发送邮件消息时，它是作为SMTP客户在运行的；当一个邮件服务器从其他邮件服务器接受邮件消息时，它是作为SMTP服务器在运行的。

 **Key words:** electronic mail（电子邮件），SMTP（Simple Mail Transfer Protocol，简单邮件传输协议）

6.4.3 FTP（文件传送协议）

FTP (File Transfer Protocol) is a protocol for transferring a file from one host to another host. The protocol dates back to 1971 (when the Internet was still an experiment), but remains enormously popular.

文件传送协议（FTP）是一个用于从一台主机到另一台主机传送文件的协议。该协议的历史可追溯到1971年（当时互联网尚处于实验之中），不过至今仍然极为流行。

In a typical FTP session, the user is sitting in front of one host (the local host) and wants to transfer files to or from a **remote** host. In order for the user to access the remote account, the user must provide a **user identification** and a password. After providing this **authorization** information, the user can transfer files from the local file system to the remote file system **and vice versa**.

一个典型的FTP会话中，用户坐在本地主机前，想把文件传送到一台远程主机或把它们从一台远程主机传送过来，该用户必须提供一个用户名和口令才能访问远程账号，给出这些身份认证信息后，它就可以在本地文件系统和远程文件系统之间传送文件了，反之亦然。

FTP uses two **parallel** TCP connections to transfer a file, a control connection and a data connection. The control connection is used for sending control information between the two hosts-information such as user identification, password, commands to change remote directory, and commands to "put" and "get" files. The data connection is used to actually send a file. Because FTP uses a separate control connection, FTP is said to send its control information **out-of-band**.

FTP使用两个并行TCP连接来传送文件，一个是控制连接，一个是数据连接。控制连接用于在客户主机和服务器主机之间发送控制信息，例如用户名和口令、改变远程目录命令、取来或放回文件命令；而数据连接用于真正发送文件。既然FTP使用一个独立的控制连接，我们说FTP是带外（out-of-band）发送控制消息的。

 **Key words:** FTP（File Transfer Protocol，文件传输协议），remote（远程），user identification（用户名），authorization（认证），and vice versa（反之亦然），parallel（并行），out-of-band（带外）

6.4.4 Internet of Things（物联网）

The **Internet of Things** refers to uniquely identifiable objects (things) and their virtual representations in an Internet-like structure. The term Internet of Things was first used by Kevin Ashton in 1999. The concept of the Internet of Things first became popular through the Auto-ID Center and related market analysts publications. **Radio-frequency identification (RFID)** is often seen as a prerequisite for the Internet of Things. If all objects of daily life were equipped with **radio tags**, they could be identified and inventoried by computers. However, unique identification of things may be achieved through other means such as **barcodes** or **2D-codes** as well.

With all objects in the world equipped with minuscule identifying devices, daily life on Earth would undergo a transformation. Companies would not run out of stock or waste products, as involved parties would know which products are required and consumed. Mislaid and stolen items would be easily tracked and located, as would the people who use them. Your ability to interact with objects could be altered remotely based on your current status and existing user agreements.

物联网是指独特的可识别对象，和它们在类似于互联网结构的虚拟表示。物联网这个词在 1999 年被凯文·阿什顿首次使用。这个概念最开始是通过 Auto-ID 中心和相关的市场分析出版物流行起来的。射频识别（RFID）往往被视为物联网的一个先决条件。假如日常生活中的所有物品都配备有电子标签，它们可以通过计算机进行识别和清点。然而，物品的唯一标识可以通过其他手段如条形码或二维码来获得。

当世界上所有的物品配备微不足道的识别装置，地球上的生活将发生转变。公司不会用完原料或浪费产品，因为作为当事部门就知道哪些产品是人们需要的。遗忘物品或者被盗物品将很容易地被跟踪和定位，因此使用它们的人也被跟踪和定位了。在当前的状态和现有的用户协议基础上，可以远程改变你和物品交互的能力。

 **Key words:** Internet of Things（物联网），RFID（Radio-frequency identification，射频识别），radio tag（电子标签），barcode（条形码），2D-code（二维码）

6.4.5 Cloud Computing（云计算）

Cloud computing refers to the delivery of computing and storage capacity as a service to a heterogeneous community of end-recipients. The name comes from the use of clouds as an abstraction for the complex infrastructure it contains in system diagrams. Cloud computing entrusts services with a user's data, software and computation over a network. It has considerable overlap with **software as a service (SaaS)**.

云计算是指把传送计算和存储的能力作为一种服务提供给不同的终端接收者。这个名字来源于在系统图中用云表示抽象的、复杂的基础设施。云计算委托基于网络的用户数据、软件和计算服务。它与软件即服务（SaaS）有相当大的重叠。

End users access cloud based applications through a Web browser or a light weight desktop or mobile app while the business software and data are stored on servers at a remote location. Proponents claim that cloud computing allows enterprises to get their applications up and running faster, with improved manageability and less maintenance, and enables IT to more rapidly adjust resources to meet fluctuating and unpredictable business demand.

终端用户通过 Web 浏览器、轻量级的台式机或移动应用程序访问基于应用的云，而商业软件和数据存储在远程位置的服务器上。支持者声称，云计算使企业的应用程序运行起来更快速，可管理性也得到了提高，维护也少，并且使 IT 能够更迅速地调整资源，以满足市场波动和不可预测的业务需求。

Cloud computing relies on sharing of resources to achieve coherence and economies of scale similar to a utility (like the **electricity grid**) over a network (typically the Internet). At the foundation of cloud computing is the broader concept of converged **infrastructure** and shared services.

云计算依赖于资源共享以实现一致性，以及类似基于网络（像典型的互联网）的公用事业（如电网）的规模经济。云计算的基础是融合了基础设施和共享服务更广泛的概念。

 **Key words:** cloud computing（云计算），SaaS（software as a service，软件即服务），electricity grid（电网），infrastructure（基础设施）

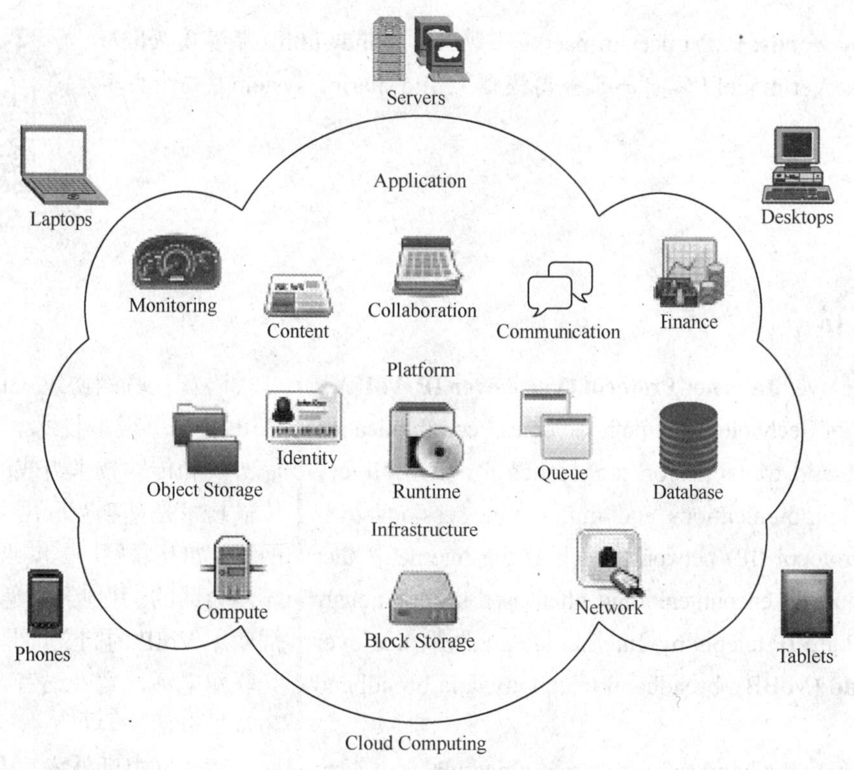

Cloud Computing Logical Diagram

6.4.6 Others（其他）

1. P2P

Peer-to-peer (P2P) computing or networking is a distributed application architecture that partitions tasks or workloads between peers. Peers are equally privileged in the application. They are said to form a peer-to-peer network of nodes.

Peers make a portion of their resources, such as processing power, disk storage or network **bandwidth**, directly available to other network participants, without the need for central coordination by servers or stable hosts. Peers are both suppliers and consumers of resources, in contrast to the traditional **client–server model** where only servers supply (send), and clients consume (receive).

The peer-to-peer application structure was popularized by **file sharing systems** like Napster.

对等计算或网络是一个分布式应用程序架构，它在端到端之间完成分区任务或者负载。在应用中每个端都是平等的，没有特权的，这样就形成一个对等节点的网络。

每个端使自己成为资源的一部分，这些资源是指如处理能力、磁盘存储或网络带宽等可以直接供其他网络参与者使用的资源，不需要服务器或固定的主机来做中央协调。相对于传统的客户—服务器模型来说，每个端都是资源的提供者，同时也是消费者。传统的客户—服务器模型中，只有唯一的服务器提供（发送），客户端消费（接收）。

对等应用结构被广泛应用于文件共享系统，如Napster。

 **Key words:** P2P（peer-to-peer，端到端），bandwidth（带宽），client—server model（客户—服务器模型），file sharing system（文件共享系统）

2. VoIP

Voice over Internet Protocol (Voice over IP, VoIP) is a family of technologies, methodologies, communication protocols, and transmission techniques for the delivery of voice communications and multimedia sessions over Internet Protocol (IP) networks, such as the Internet. Other terms frequently encountered and often used synonymously with VoIP are IP telephony, Internet telephony, voice over **broadband** (VoBB), broadband telephony, and broadband phone.

Internet telephony refers to communications services-Voice, fax, SMS, and/or voice-messaging applications-that are transported via the Internet, rather than the **public switched telephone network (PSTN)**. The steps involved in originating a VoIP telephone call are signaling and media channel setup, digitization of the analog voice signal, encoding, packetization, and transmission as Internet Protocol (IP) packets over a **packet-switched network**. On the receiving side, similar steps (usually in the reverse order) such as reception of the IP packets, decoding of the packets and digital-to-analog conversion reproduce the original voice stream.

语音互联网协议（IP 语音，VoIP）是由很多技术、方法、通信协议和传输技术组成的，其中，传输技术是基于互联网协议（IP）网络，如互联网，提供语音通信多媒体会话的。其他经常遇到的和经常使用 VoIP 的同义词是 IP 电话、互联网电话、宽带语音（VoBB）和宽带电话。

互联网电话技术，是指语音、传真、短信和语音消息这些通信服务应用通过互联网传输，而不是公共电话交换网（PSTN）。参与发起一个 VoIP 电话的步骤是设置信号和媒体通道，模拟语音信号的数字化，编码，打包，并作为一个在分组交换网络上的互联网协议（IP）数据包传输。在接收端，类似的步骤，如接收 IP 数据包（通常是相反的顺序），解码包和数字—模拟转换重现原始的语音流。

 **Key words:** Voice over IP, VoIP（Voice over Internet Protocol，IP 语音），broadband（宽带），PSTN（public switched telephone network，公共电话交换网），packet-switched network（分组交换网）

6.5 Computer Network Safety（计算机网络安全）

6.5.1 Firewall（防火墙）

As the name suggests, a **firewall** protects data from the outside world. A firewall can be a software program or a hardware device. A firewall is a popular security mechanism for networks. A firewall is a simple router implemented with a special program. This unit is placed between hosts of a certain network and the outside world. The security issues faced by a smaller network like the one used at home is similar to larger networks. A firewall is used to protect the network from unwanted Web sites and potential **hackers**.

顾名思义，防火墙保护与外部世界往来的数据，它可以是一个软件程序或硬件设备。防火墙是一种流行的网络安全机制，是一个带有特殊程序的简单路由器，它被放置在某个网络主机与外部世界之间。在家使用一个较小网络所面临的安全问题和大型网络所面临的安全问题是类似的，防火墙常用来保护网络避免有害网站和潜在的黑客的攻击。

A firewall is placed on the link between a network router and the Internet or between a user and a router. The objective of such a configuration is to monitor and filter packets coming from unknown sources. Consequently, hackers do not have access to penetrate through a system if a firewall protects the system. In addition, the firewall can control how a network works with an Internet connection. A firewall can also be used to control data traffic.

防火墙被放置在网络路由器和互联网之间，或者放置在用户和路由器之间，这种构造的目的是监测和过滤来源不明的数据包，因此，如果防火墙保护该网络，黑客不能进入该网络中。另外，防火墙可以控制与互联网连接的网络如何工作，还可以用于控制数据传输。

A firewall controls the flow of traffic by one of the following three methods. The first method is packet filtering. Apart from forwarding packets between networks, a firewall filters those packets that pass through. If packets can get through the filter, they reach their destinations; otherwise, they are discarded. The second method is that a firewall filters packets based on the source IP address. This filtering is helpful when a host has to be protected from any unwanted external packets. The third method, denial of service. This method controls the number of packets entering a network.

防火墙通过以下 3 种方法控制数据的流量：第 1 种方法是包过滤，除了网络间转发的数据包，数据包都通过防火墙进行包过滤。如果数据包可以通过过滤器，它们就可以到达目的地，否则，它们将被丢弃。第 2 种方法是，防火墙基于源 IP 地址过滤数据包。当主机必须保护任何不需要的外部数据包时，这种过滤是有帮助的。第 3 种方法是拒绝服务攻击。这种方法是控制进入网络的数据包的数量。

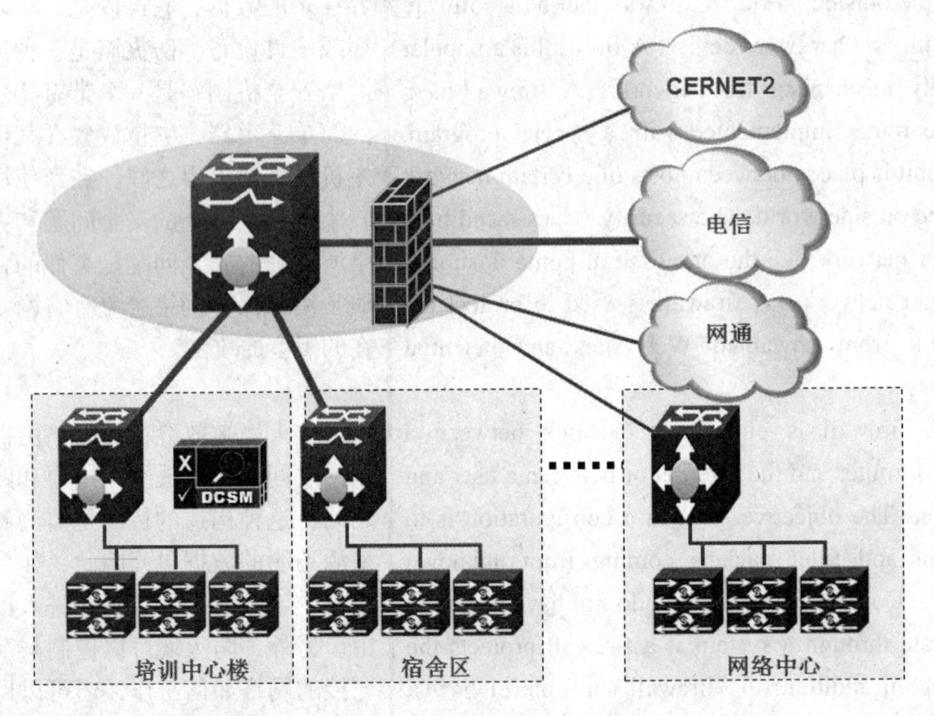

Key words: firewall（防火墙），hacker（黑客）

6.5.2 IDS（入侵检测系统）

An **intrusion detection system (IDS)** is a device (or application) that monitors network and/or system activities for malicious activities or policy violations.

Intrusion detection is the process of monitoring the events occurring in a computer system or network and analyzing them for signs of possible incidents, which are violations or imminent threats of violation of computer security policies, acceptable use policies, or standard security practices. Intrusion prevention is the process of performing intrusion detection and attempting to stop detected possible incidents. **Intrusion detection and prevention systems (IDPS)** are primarily focused on identifying possible incidents,

入侵检测系统（IDS）是一个设备（或应用），它可以监测网络或系统的恶意活动或违反政策的活动。

入侵检测是监测在计算机系统或网络的事件，并且分析可能发生的事件，该事件是侵犯或即将发生违反计算机安全政策、可接受的使用政策或标准的安全常规这样的威胁。入侵防御是执行入侵检测，并试图阻止检测到的可能发生的事件的过程。入侵检测和防御系统（IDPS）主要是识别可能发生的事件，记录事件的信息，试图制止事件的发生，并将它们

logging information about them, attempting to stop them, and reporting them to security administrators. In addition, organizations use IDPSs for other purposes, such as identifying problems with security policies, documenting existing threats, and deterring individuals from violating security policies. IDPSs have become a necessary addition to the security infrastructure of nearly every organization.

报告给安全管理员。此外，使用 IDPS 的组织还有其他的意图，例如识别安全政策问题，记录现有的威胁，以及阻止个人违反安全政策。IDPS 几乎已经成为每个组织必要的安全基础设施的补充。

IDPSs typically record information related to observed events, notify security administrators of important observed events, and produce reports. Many IDPSs can also respond to a detected threat by attempting to prevent it from succeeding. They use several response techniques, which involve the IDPS stopping the attack itself, changing the security environment (e.g., reconfiguring a firewall), or changing the attack's content.

IDPS 记录与观察与事件相关的代表性的信息，通知重要事件的安全管理人员，并产生报告。许多 IDPS 也可以对成功阻止威胁的检测做出回应，这种 IDPS 使用多种响应技术，其中包括 IDPS 停止自身攻击、不断变化的安全环境（例如重新配置防火墙）或者改变攻击的内容。

 **Key words:** IDS（intrusion detection system，入侵检测系统），IDPS（intrusion detection and prevention systems，入侵检测和防御系统）

6.6 Situation Dialogue（情境对话）

Xiao Ming and Li Liang are good friends. Li Liang is a computer expert who often buys things online.

小明和李亮是好朋友。李亮是计算机高手，经常在网上买东西。

Xiao Ming: More and more people like on-line shopping.

小明：现在喜欢网上购物的人越来越多了。

Li Liang: Yes, the price is lower online.

李亮：是呀，网上的东西价格比较便宜。

Xiao Ming: I saw a digital camera online.

小明：我也在网上看中一款数码相机。

Li Liang: Are you ready to do shopping online?

李亮：你也准备网上购物？

Xiao Ming: Yes, but I do not know how to do it, can you show me?

小明：是的，但是我不知道怎么做，你能教教我吗？

Li Liang: OK. There are many fishing sites on the network, so firstly, selecting the site which you can trust is very important .

李亮：好的。网络上有很多钓鱼网站，选择可以信任的网站是非常重要的。

Xiao Ming: What sites?

小明：有哪些网站呢？

Li Liang: To buy a digital camera, there are Jingdong, Newegg, Taobao, eBay, and so on.

李亮：买数码相机的话，有京东、新蛋、淘宝、易趣等。

Xiao Ming: Can it be traded after choosing the right products?

小明：选好产品后就可以进行交易了？

Li Liang: Yes, select a digital camera, pay, receive digital camera. It's so simple.

李亮：是的，选数码相机，付钱，收到数码相机，就是这么简单。

Xiao Ming: No wonder everyone likes online shopping.

小明：难怪大家都喜欢网上购物。

Li Liang: Although simple, many techniques are used. Take the network side, there are HTTP, DNS, network device configuration, the development of standards and protocols, and so on.

李亮：虽然简单，但用到的技术很多。就拿网络方面来说，主要有 HTTP、DNS、网络设备配置、标准和协议的制定等。

Xiao Ming: It sounds very complicated.

小明：听起来很复杂。

Li Liang: Of course, online chatting and watching movies use many networking technologies.

李亮：当然，网上聊天，看电影，这些都用到很多的网络技术。

Xiao Ming: Now I turn the computer on, How do you think the digital camera I chosed?

小明：我把电脑打开，你看我选的数码相机怎么样？

Li Liang: OK.

李亮：好的。

Situation Dialogue

6.7 Reading and Compacting（对照阅读）

What are Computer Viruses?

Computer viruses are programs written by "mean"

什么是计算机病毒？

计算机病毒是不良之人撰写出

people. These virus programs are placed into a commonly used program so that program will run the attached virus program as it boots, therefore, it is said that the virus "infects" the executable file or program. Executable files include Macintosh "system files" (such as system extensions, INITs and control panels) and application programs (such as word processing programs and spreadsheet programs). Viruses work the same ways in Windows or DOS machines by infecting zip or exe files.

的程序。这些病毒程序被放置在一个常用的程序中，所以此（正常的）程序在启动时就会运行附加其上的病毒程序，因此，我们说该病毒"感染"可执行文件或程序。可执行文件包括Macintosh"系统文件"（如系统扩展、初始化和控制面板）和应用程序（如文字处理程序和电子表格程序）。病毒在Windows或DOS机器上感染zip或exe文件采用相同的工作方式。

A virus is inactive until you execute an infected program or application or start your computer from a disk that has infected system files. Once a virus is active, it loads into your computer's memory and may save itself to your hard drive or copies itself to applications or system files on disks you use.

病毒处于非活动状态，直到你执行受感染的应用程序或从磁盘启动已经感染的计算机系统，文件才被激活。病毒一旦激活，它将加载到你的计算机内存中，并可能保存到你的硬盘驱动器或自身复制到磁盘上的应用程序或系统文件中。

Some viruses are programmed specifically to damage the data on your computer by corrupting programs, deleting files, or even erasing your entire hard drive. Many viruses do nothing more than display a message or make sounds/verbal comments at a certain time or a programming event after replicating themselves to be picked up by other users one way or another. Other viruses make your computer's system behave erratically or crash frequently. Sadly many people who have problems or frequent crashes using their computers do not realize that they have a virus and live with the inconveniences.

一些病毒是为了损害你的计算机上的数据专门设计的程序，它们通过破坏程序、删除文件，甚至清除你的整个硬盘驱动器。许多病毒只不过显示一条消息，或发出提示音/口头说明就自我复制至其他用户，自我复制是在某一特定的时间或者程序事件后执行的。其他病毒使计算机系统的行为不正常或频繁崩溃。可悲的是，很多使用有问题的计算机用户没有意识到，是病毒给他们的生活造成了不便。

What Viruses Don't Do!

什么是病毒做不了的！

Computer viruses can not infect write protected disks or infect written documents. Viruses do not infect compressed files, unless the file was infected prior to the compression. Compressed files are programs or files with its common characters, etc. removed to take up less space on a disk. Viruses do not infect computer hardware, such as monitors or computer chips; they only infect software.

计算机病毒无法感染写保护的磁盘或书面文件。病毒不感染压缩的文件，除非文件是在被压缩之前被感染的，压缩文件程序具有的共同特征文件占用的磁盘空间较少。病毒不感染计算机硬件，如显示器或计算机芯片，它们只感染软件。

In addition, Macintosh viruses do not infect DOS/Window computer software and vice versa. For example, the Melissa virus incident of late 1998 and the ILOVEYOU virus of 2000 worked only on Window based machines and could not operate on Macintosh computers.

此外，Macintosh病毒不感染DOS/窗口计算机软件，反之亦然。例如，1998年底的Melissa病毒和2000年ILOVEYOU病毒只能在基于窗口工作的机器上工作，不能在Macintosh计算机上运行。

One further note is viruses do not necessarily let you know they are present in your machine, even after being destructive. If your computer is not operating properly, it is a good practice to check for viruses with a current "virus checking" program.

还有一项值得注意的是，病毒并不一定让你知道它们目前是在你的机器上，虽然它们之后会有破坏性。如果您的计算机没有正常运行，用流行的查毒软件检查病毒是一个很好的做法。

How do Viruses Spread?

病毒如何传播？

Viruses begin to work and spread when you start up the program or application of which the virus is present. For example, a word processing program that contains a virus will place the virus in memory every time the word processing program is run.

当你启动带病毒的程序或应用程序时，病毒开始工作并传播。例如，包含一个病毒的文字处理程序将会把该病毒放在内存中，随着每次文字处理程序的运行而传播病毒。

Once in memory, one of a number of things can happen. The virus may be programmed to attach to other applications, disks or folders. It may infect a network if given the opportunity.

一旦病毒进入内存，将发生许多事情。该病毒可被编程附加到其他应用程序、磁盘或文件夹，如果有机会它还可能感染网络。

Viruses behave in different ways. Some viruses stay active only when the application it is part of is running. Turn the computer off and the virus is inactive. Other viruses will operate every time you turn on your computer after infecting a system file or network.

病毒的传播方式不同。有些病毒只是当应用程序运行它所在的那一部分时保持激活状态，关闭计算机时病毒处于不活动状态。其他病毒则在每次打开计算机后，感染系统文件或网络。

6.8 术语简介

1. TCP/IP Model：Transmission Control Protocol/Internet Protocol 模型，TCP/IP 起源于20世纪60年代末美国政府资助的一个网络分组交换研究项目，TCP/IP是发展至今最成功的通信协议，它被用于当今所构筑的最大的开放式网络系统互联网之上。TCP和IP是两个独立且紧密结合的协议，负责管理和引导数据报文在互联网上的传输。二者使用专门的报文头定义每个报文的内容。

2. OSI Model：开放系统互连参考模型（Open System Interconnection Reference Model，OSI/RM），它是由国际标准化组织（International Standard Organization，ISO）提出的一个

网络系统互连模型。

3．HTTP：超文本传输协议，是 Hyper Text Transfer Protocol 的缩写。HTTP 定义了信息如何被格式化、如何被传输，以及在各种命令下服务器和浏览器所采取的响应。

4．WWW：WWW 是 World Wide Web 的简称，译为万维网，是指在互联网上以超文本为基础形成的信息网。

5．FTP：文件传输协议，是 File Transfer Protocol 的缩写。它是互联网上使用非常广泛的一种通信协议，是计算机网络上主机之间传送文件的一种服务协议。

6．E-mail：是 Electronic Mail 的缩写，即电子邮件。E-mail 是一种常用的互联网服务。就是利用计算机网络交换的电子媒体信件。

Exercises（练习）

1．Match the explanations in Column B with words and expressions in Column A.（搭配每组中同意义的词或短语）

A	B	A	B
交换机	IP	超文本传输控制协议	Network Safety
路由器	IDPS	电子邮件	Firewall
组件	Switch	文件传输协议	E-mail
传输控制协议	Router	防火墙	IDS
互联网协议	Component	入侵检测系统	FTP
万维网	TCP	网络安全	OSI
入侵检测和防御系统	WWW	开放系统互联	HTTP

2．Choose the proper words to fill in the blanks.（选词填空）

A LAN is used for communications in a small community in which resources, such as (　　), software, and servers, are shared. Each device connected to a LAN has a unique (　　). Two or more LANs of the same type can also be connected to (　　) data frames among multiple users of other local area networks. In LANs, (　　) are additional headers appended for local routing. These new-looking packets are known as (　　).

> Words to be chosen from:（可选词）
> Packets，address，frames，printers，forward

3．Translate the following English into Chinese.（将下面英语句子翻译为中文）

（1）The Internet is a global system of interconnected computer networks that use the standard Internet Protocol Suite (TCP/IP) to serve billions of users worldwide.

____

（2）As with most application-layer protocols, SMTP has two sides: a client side, which executes on the sender's mail server, and a server side, which executes on the recipient's mail server.

____

（3）In the star topology, all users are directly connected to a central user through two unidirectional links: one for uplink and the other for downlink.

____

（4）A computer network is a group of computers that are connected to each other for the purpose of communication.

____

（5）WANs are used to connect LANs and other types of networks together, so that users and computers in one location can communicate with users and computers in other locations.

____

附文6: Reading Material（阅读材料）

1. Varieties of Ethernet

Fast Ethernet

- 100BASE-T: A term for any of the three standards for 100 Mbit/s Ethernet over twisted pair cables, includes 100BASE-TX, 100BASE-T4 and 100BASE-T2. As of 2009, 100BASE-TX has totally dominated the market, and is often considered to be synonymous with 100BASE-T in informal usage.
- 100BASE-TX: 100 Mbit/s Ethernet over Category 5 cables (using two out of four pairs). Similar star-shaped configuration to 10BASE-T.
- 100BASE-T4: 100 Mbit/s Ethernet over Category 3 cables (as used for 10BASE-T installations). Uses all four pairs in the cable, and is limited to half-duplex. Now obsolete, as Category 5 cables are the norm.
- 100BASE-T2: 100 Mbit/s Ethernet over Category 3 cables. Uses only two pairs, and supports full-duplex. It is functionally equivalent to 100BASE-TX, but supports old cable. No products supporting this standard was ever manufactured.
- 100BASE-FX: 100 Mbit/s Ethernet over fiber.

Gigabit Ethernet

- 1000BASE-T: 1 Gbit/s over unshielded twisted pair copper cabling (at least Category 5 cables, with Category 5e strongly recommended).
- 1000BASE-SX: 1 Gbit/s over short range multi-mode fiber.
- 1000BASE-LX: 1 Gbit/s over long range single-mode fiber.
- 1000BASE-CX: A short-haul solution (up to 25 m) for running 1 Gbit/s Ethernet over special copper cable. Predates 1000BASE-T, and now obsolete.

10-gigabit Ethernet

The 10 gigabit Ethernet family of standards encompasses media types for single-mode fiber (long haul), multi-mode fiber (up to 300 m), copper backplane (up to 1 m) and copper twisted pair (up to 100 m). It was first standardised as IEEE Std 802.3ae-2002, but is now included in IEEE Std 802.3-2008.

- 10GBASE-SR: designed to support short distances over deployed multi-mode fiber cabling, it has a range of between 26 m and 82 m depending on cable type. It also supports 300 m operation over a new 2000 MHz·km multi-mode fiber.
- 10GBASE-LX4: uses wavelength division multiplexing to support ranges of between 240 m and 300 m over deployed multi-mode cabling. Also supports 10 km over single-mode fiber.
- 10GBASE-LR and 10GBASE-ER: these standards support 10 km and 40 km respectively over single-mode fiber.
- 10GBASE-SW, 10GBASE-LW and 10GBASE-EW. These varieties use the WAN PHY, designed to interoperate with OC-192 / STM-64 SONET/SDH equipment. They correspond at the physical layer to 10GBASE-SR, 10GBASE-LR, and 10GBASE-ER respectively, and hence use the same types of fiber and support the same distances. (There is no WAN PHY standard corresponding to 10GBASE-LX4.)
- 10GBASE-T: designed to support copper twisted pair was specified by the IEEE Std 802.3an-2006 which has been incorporated into the IEEE Std 802.3-2008.

2. ARPANET

The ARPANET (Advanced Research Projects Agency Network) created by the ARPA of the United States Department of Defense during the Cold War, was the world's first operational packet switching network, and the predecessor of the global Internet.

Packet switching, now the dominant basis for both data and voice communication worldwide, was a new and important concept in data communication. Previously, data communication was based on the idea of circuit switching, as in the old typical telephone circuit, where a dedicated circuit is tied up for the duration of the call and communication is only possible with the single party on the other end of the circuit.

With packet switching, a system could use one communication link to communicate with more than one machine by disassembling data into datagrams, then gather these as packets. Not only could the link be shared (much as a single post box can be used to post letters to different destinations), but each packet could be routed independently of other packets.

A form of packet switching designed by Lincoln Laboratory scientist Lawrence Roberts underlay the design of ARPANET.

Background of ARPANET

The earliest ideas of a computer network intended to allow general communication between users of various computers were formulated by J.C.R. Licklider of Bolt, Beranek and Newman (BBN) in August 1962, in a series of memos discussing his "Intergalactic Computer Network" concept. These ideas contained almost everything that the Internet is today.

In October 1963, Licklider was appointed head of the Behavioral Sciences and Command and Control programs at ARPA (as it was then called), the United States Department of Defense Advanced Research Projects Agency. He then convinced Ivan Sutherland and Bob Taylor that this was a very important concept, although he left ARPANET before any actual work on his vision was performed.

ARPA and Taylor continued to be interested in creating a computer communication network, in part to allow ARPA-sponsored researchers in various locations to use various computers which ARPA was providing, and in part to make new software and other results widely available quickly. Taylor had three different terminals in his office, connected to three different computers which ARPA was funding: one for the SDC Q-32 in Santa Monica, one for Project Genie at the University of California, Berkeley, and one for Multics at MIT. Taylor later recalled:

"For each of these three terminals, I had three different sets of user commands. So if I was talking online with someone at S.D.C. and I wanted to talk to someone I knew at Berkeley or M.I.T. about this, I had to get up from the S.D.C. terminal, go over and log into the other terminal and get in touch with them. I said, oh, man, it's obvious what to do: If you have these three terminals, there ought to be one terminal that goes anywhere you want to go. That idea is the ARPANET".

Somewhat contemporaneously, a number of people had (mostly independently) worked out various aspects of what later became known as "packet switching", with the first public demonstration being made by the UK's National Physical Laboratory (NPL) on 5 August 5, 1968. The people who created the ARPANET would eventually draw on all these different sources.

3. Hotmail

In December 1995, Sabeer Bhatia and Jack Smith visited the Internet venture capitalist Draper Fisher Jurvetson and proposed developing a free Web-based E-mail system. The idea was to give a free E-mail account to anyone who wanted one, and to make the accounts accessible from the Web. With Web-based E-mail, anyone with access to the Web—say, from a school or community library—could read and send E-mails. Furthermore, web-based E-mail would offer great mobility to its subscribers. In exchange for 15 percent of the company, Draper Fisher Jurvetson financed Bhatia and Smith, who formed a company called Hotmail. With three full-time people and 12 ~ 14 part-time people who worked for stock options, they were able to develop and launch the service in July 1996. Within a month after launch they had 100 000 subscribers. The number of subscribers continued to grow rapidly, with all of their subscribers being exposed to advertising banners while reading their E-mail. In December 1997, less than 18 months after launching the service, Hotmail had over 12 million subscribers and was acquired by Microsoft, reportedly for $400 million dollars.

The success of Hotmail is often attributed to its "first-mover advantage" and to the inherent "viral marketing" of E-mail. Hotmail had a first-mover advantage because it was the first company to offer Web-based E-mail. Other companies, of course, copied Hotmail's idea, but Hotmail had a six-month lead on them. The coveted first-mover advantage is obtained by having an original idea, and then developing quickly and secretly. A service or a product is said to have viral marketing if it markets itself. E-mail is a classic example of a service with viral marketing—the sender sends a message to one or more recipients, and then all the recipients become aware of the service. Hotmail demonstrated that the combination of first-mover advantage and viral marketing can produce a killer application. Perhaps some of the students reading this book will be among the new entrepreneurs who conceive and develop first-mover Internet services with inherent viral marketing.

Chapter 7 IT Workplace English（IT 职场英语）

教学要求

掌握常见职场英语词汇；撰写简单的英语简历；制作英文 PowerPoint；提高英语面试口语能力。

教学内容

职场英语口语；英语自荐书；常用的专业术语。

教学提示

创建模拟场景，锻炼口语能力。

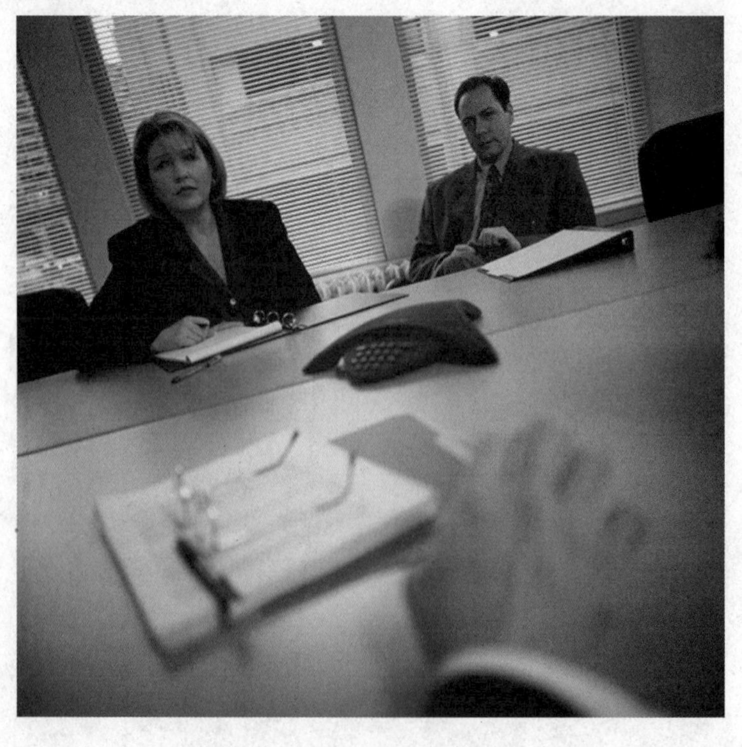

Cultural Background（文化背景）

In the States, there are many ways to find a job. Many American universities hold annual job fairs, which is a special period of time for companies send representatives to **campuses** to recruit. This is the best opportunity for those who will graduate soon to find a job. If you have a chance to attend a job fair, you may bring copies of your resume so that you can provide to your potential employers.

You can also approach jobs from newspaper **employment** ads section. Companies may advertise in newspaper and give a description about the position and the skills required for it. In addition, job seekers may go to employment agencies. Companies will make decisions based on whether candidate's ability meet requirements.

在美国求职的途径很多，许多美国大学每年都举办人才交流会。人才交流会是众多公司派代表到大学招聘新雇员的特殊形式，这对那些即将毕业的学生来说是求职的最佳机会。若有机会参加这样的交流会，需要准备数份个人简历投给有可能成为雇主的人。

在美国还可以利用报纸的招聘广告栏来求职。公司可以在报纸上刊登招聘广告，说明工作性质及所需技能。另外，求职者也可以去就业机构，公司只需依据求职者是否具有符合工作条件的能力来决定是否雇用。

Key words: campus（校园），employment（雇用、职业、工作）

7.1 IT Workplace Spoken English （IT 职场英语口语）

7.1.1 Office Communication 1（办公室交流 1）

A：Is your computer fast enough?

B：Oh my god, my computer is **unbelievably** slow.

A：你的计算机运行速度快吗？

B：哦，天啊！我的计算机速度慢得难以置信。

A：Why? It can't be that bad. Didn't human resource just **issue** you a new computer last month? If it is new, the **processor** should be pretty decent, it shouldn't be that slow.

A：为什么？它不应该那么糟糕吧，公司人力资源部不是上个月才给你发了一台新的计算机吗？如果是新的，处理器应该不错，它不该那么慢的。

B：It's slower than my dead grandmother! The computer they gave me last month, but it was already 3 years old.

B：它比我去世的祖母动作还慢！他们上个月给我一台计算机，但是那台计算机已经用3年了。

A：Is that right? I sympathize with you. How much **RAM** does it have?

A：是这样吗？我对此深表同情。它的随机存储器有多大？

B：I don't know, but I know my computer definitely does not have enough memory.

B：我不确定，但是我知道计算机一定没有足够的内存。

A：Do you have to save some of the songs on your computer? I want to copy some of the songs from your computer to my iPod. I usually like to update my iPod every few days with some new **tunes** to listen to on my **commute**.

A：你的计算机里存有歌曲吗？我想从你的计算机上拷贝一些到我的 iPod上。我通常每隔几天就更新一些歌曲，好在我上下班的途中听。

B：I have some songs on my computer, How many songs can you put on your iPod?

B：我的计算机上有一些歌曲，你的 iPod 能存多少首歌曲？

A：It's got 8 **gigs**, so I can put a lot of **stuff** in it.

A：它有 8G 的空间，我能放很多东西在里边。

Key words: unbelievably（难以置信地），issue（分发、给），processor（处理器），RAM（随机存储器），tune（曲子），commute（上下班两地之间坐车往返），gig（内存空间，G），stuff（物品、东西）

7.1.2 Office Communication 2（办公室交流2）

A：Do you want me to print you a copy for my **report**?

A：要我帮你打印一份报告给你吗？

B：No, just E-mail it to me, if you can.

B：不，如果可以的话，发邮件给我就可以了。

A：Good, that way you can just make your corrections on line.

A：好的，那你可以在网上进行修改。

B：Yes, and we can save paper.

B：是的，并且我们可以省下纸张。

A：By the way, what is your **E-mail address**?

A：顺便问一下，你的邮箱地址是多少？

B：My E-mail address is ABC@163.com.

A：Good, I'll send your mail. Does your office have E-mail?

B：Yes. I couldn't live without it，you can use E-mail to send messages to me.

B：我的邮箱地址是 ABC@163. com。

A：好，我会发邮件给你的，你的办公室能收邮件吗？

B：可以的，我随时都在用它，你可以用电子邮件把信息传给我。

 **Key words:** report（报告），E-mail address（电子邮件地址）

7.2 Introduce Myself（自我介绍）

A **resume** is a self-promotional **document**, which is supposed to present you in the best way and get you a **interview** chance. An effective resume majorly tells who you are, your capability, and your recent education and training experience. It not just present your past jobs, but also need include how you performed and what you accomplished in those jobs. A good resume can help to **predict** your career path in the future.

简历是一种用来介绍自己并以此来吸引他人眼光的文件，它可以帮助简历制作者获得一个工作面试的机会。一个有效的简历应重点介绍你是谁、你的能力，以及你的最新教育经历或培训经验。它不仅包括你过去所从事的工作，还应说明你在过去的工作表现如何，完成了哪些事情。一份好的简历，可以帮你预估出将来所要从事的工作。

Writing a resume is a **challenge** for students even for people who have work experience. When a college new graduate write a resume, the biggest problem is the lack of relevant experience. They can barely compete with a experienced worker, because without work experience, employer barely can evaluate the candidate's ability. So as a new graduate, you must analyze and **assess** your **relevant** skills as well as develop a list of relevant **achievements** that you can put in your resume. This is not the time to be modest.

写一份简历对学生来说是一项挑战，甚至对有经验的工作者也是这样。简历写作最大的问题是，大多数大学应届毕业生拥有的相关工作经验很少。他们几乎无法和一个有经验的求职者相比。因为没有工作经历，任何雇主都很难评估候选人的工作能力。所以作为一名应届毕业生，你必须分析和评估自己拥有的相关技能以及要在简历里写的一系列相关成就。这不是谦虚的时候。

Key words: resume（简历）, document（文件）, interview（接见、会见、面试）, predict（预测）, challenge（挑战）, assess（估定价值、确定金额）, relevant（有关的、有意义的）, achievement（成就）

7.3 Career Ability（职场能力）

7.3.1 How to Make a Good PPT Presentation （如何做一个好的 PPT 简报）

Making a good PowerPoint **presentation** is not hard, but it does require some **forethought**. The simpler your presentation is, the more likely your audience will be to understand and remember the information you present.

制作一个好的 PPT 简报并不难，但它需要花一些心思。你的 PPT 做得越简单，你的观众越有可能明白和记住你所要传达的信息。

No.1, put the main points of your speech **outline** into your PowerPoint presentation. Include only one idea per slide. As J. L. Doumont **suggests** in the September 2004 issue of "Technical Communication", use as little text as possible so that the slides do not draw the audience's attention away from the **speaker**.

第一，把你的主要发言提纲放入你的演示文稿。每张幻灯片只包括一个主题。杜蒙·特路在 2004 年 9 月的《技术交流》上建议，使用尽可能少的文字，否则，人们的注意力都在幻灯片上了，而不会注意发言者了。

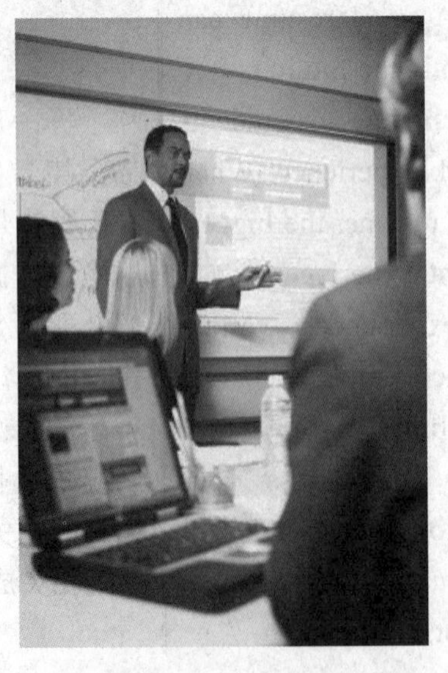

Key words: presentation（描述、陈述、介绍、发言），forethought（考虑、深谋远虑），outline（大纲、概要），suggest（建议），speaker（发言者）

No.2, design your slides. Your presentation will be more **effective** if you steer clear of templates. Avoid designing a presentation with so many colors and effects that the audience remembers the design more than the information in your speech. **According to** Doumont, "presenters can usefully develop a first design in black and white, then add color in light touches, for **emphasis** or **identification**". Using mainly black and white will also make the slides easier for the audience to read.

第二，设计你的幻灯片。如果你不使用模板，你的展示将更加有效。不要设计使用太多的颜色和效果，这样会导致在你的演讲中观众记住的设计比信息多。据杜蒙·特路所说，"展示者可以先只用黑色和白色做最初的设计，然后再添加光线和色彩，用于强调或识别"。用黑色和白色也将使观众更容易阅读幻灯片。

No.3, insert graphics. **Graphics** can help **audience** members remember your points, but unrelated graphics will hurt your presentation. The graphics must be relevant to the subject and should not **crowd** the slides; otherwise you risk confusing the audience.

第三，插入图形。图形可以帮助观众记住你的观点，但无关的图形会影响你的展示。图形必须与主题相关，不应塞满幻灯片，否则有使观众混淆的风险。

Key words: effective（有效的），according to（据……所说），emphasis（重点、强调），identification（识别、认同），graphic（图形），audience（观众），crowd（塞满）

Gill Sans: Gill Sans 字体是 Monotype 公司 1928 年设计的一种字体，Gill Sans 字体被指定为第 28 届雅典奥运会官方打印字体。大概这种字体的设计师吉尔先生（Eric Gill）也预想不到，无论潮流如何变化、技术如何发展，他设计的字体始终显示出如此强大的生命力，而且 Gill Sans 字体还与奥运会的要求非常吻合。实际上，设计 Gill Sans 字体的吉尔先生本人既是一个书法家，也是一位雕刻家。

No.4, choose **fonts**. A good sans serif font, such as Gill Sans, will make your slides easier for the audience to read. Use a combination of **capital and lowercase letters**. As noted by Michael Alley and Kathryn Neeley in the April 2005 issue of "Technical Communication", text written in only capital letters is harder for the audience to read.

第四，选择字体。一个好的无衬线字体，如 Gill Sans 字体，将使你的幻灯片更容易为读者所阅读。结合使用大小写英文字母。正如迈克尔·安里和凯瑟琳·尼利在 2005 年 4 月的《技术交流》上所说的，全部大写的书面文字是难以被观众阅读的。

No.5, keep your margins wide. Instead of cluttering each slide with too much information, leave ample room around the text and graphics. Keep in mind the goal of creating slides that are easy for the audience to read and understand quickly.

第五，保持边距。不要用太多的信息塞满每张幻灯片，在文本和图形周围留出足够的空间。记住你的目标：做出让观众易于阅读并能快速理解的幻灯片。

No.6, **proofread**. Make sure you have no **grammatical** or spelling errors. It only takes a few minutes to proofread so take the time!

第六，校对。确保没有语法和拼写错误。校对只需要几分钟，所以请抽出时间去做！

No.7, have a simple **conclusion**. Make it similar to your title page. Take the time to put in the conclusion slide. This is often forgotten many times but it is important for your audience and you. It can be used as a cue or transition into the next presentation or questioning. Don't forget to thank your audience at the end!

第七，得出一个简单的结论，类似于你的标题。花些时间放在幻灯片的结论上。结论常常会被遗忘，但对你的观众和你是很重要的。它可以被用来作为线索或是下一个内容的过渡。最后不要忘了感谢你的观众！

 **Key words:** font（字体、字形），capital and lowercase letters（英文大小写字母），proofread（校对），grammatical（语法上的），conclusion（结论）

7.3.2 How to Use E-mail Etiquette in the Workplace（在工作场所使用电子邮件的礼仪）

Expressing yourself through E-mail is helpful and necessary. But it is important to know the audience you are corresponding with. Read on to learn how to use E-mail etiquette in the workplace.

通过电子邮件来表达自己是有益和必要的。但重要的是要了解你要交流的对象。下面是一些在工作场所使用电子邮件的礼仪。

Try to familiarize yourself with standard business-letter composition. Use basic business **salutations**, such as the word "dear," as opposed to using the word "Hi". Use the person's first name if you are familiar with him and his department. If you are not familiar with the person, use his full name or last name. Try to save the salutation "Hi" for people you know on a friendly basis.

尽量让自己熟悉商务信件的格式。使用基本业务称呼，如"亲爱的"，而不是用"嗨"这个词。如果你跟他和他的部门很熟，可以直呼他的名字。如果你对他不熟悉，可以称呼他的全名或姓。和你熟悉的人在友好的基础上可以问候"嗨"。

Create a **businesslike** tone in the body of your E-mail. After a brief introduction, fill the body of your E-mail with clearly thought-out sentences. This will help the recipient understand exactly what actions need to be taken when working on a project, or what questions need to be answered.

在你的电子邮件正文里创造一个务实的基调。在一个简短的介绍后，用清晰的句子来写你的邮件内容。这有助于收信人明白在一个项目中工作需要采取的行动，或者什么样的问题需要回答。

Apply the spell-check and punctuation functions **to** your entire E-mail. Spelling and punctuation are **essential to** proper E-mail etiquette. The worst thing you can do is send an E-mail filled with misspellings and incorrect punctuation to your superiors.

把拼写检查功能和标点符号功能应用于整个电子邮件。正确拼写和标点符号的正确使用在规范的电子邮件中是必不可少的。你所做的最糟糕的事情就是向你的上级发送一封充满错误拼写和错误标点的电子邮件。

Familiarize yourself with the terms and usage of Cc: and Bcc:. "Cc:" means carbon copy and "Bcc:" means blind carbon copy. When you want a third party to see your E-mail, you Cc: them. This lets the **recipient** of the E-mail know another party has been included in the viewing of this E-mail. Bcc: allows you the option of **permitting** the third party to view this E-mail without the recipient's knowledge.

熟悉Cc和Bcc的使用。"Cc:"的意思是"抄送"，而"Bcc:"的意思是"密件抄送"。如果你想让第三方去看你的邮件，你可以抄送他们。这样邮件的收件人可以知道电子邮件可以被另一方看到。Bcc:在收件人不知情的情况下让第三方能查看电子邮件。

Key words: businesslike（有效的、务实的），apply to（把……应用于），essential to（对……是必不可少的），recipient（接受者），permit（允许）

7.4 Situation Dialogue（情境对话）

Miss Li: Our advertisement says English competence is a key requirement of this position. Then how do you think of your proficiency in written and spoken English.

Mr. Zhang: I have learned English for 10 years, and My spoken English is fairly good enough to express myself fluently.

Miss Li: Have you obtained any certificate of technical qualifications?

Mr. Zhang: Yes. I have received a Computer Operation's Qualification Certificate.

Miss Li: What special skills do you have, can you tell me?

Mr. Zhang: I have experience in computer operation , proficiency in Microsoft Windows , Microsoft Word and Microsoft Excel.

Miss Li: What computer languages have you learned?

Mr. Zhang: I have learned C.

Miss Li: Do you have any experience in programming?

Mr. Zhang: Yes. I have. I used to work as a part-time programmer in Intel Company.

Miss Li: What courses did you take?

Mr. Zhang: The courses I have taken include Computer Science, Computer Software and so on.

Miss Li: How do you get along with others?

Mr. Zhang: I get on well with others.

Miss Li: Why do you want to work for us?

Mr. Zhang: Your company is a leader in the field of electronics, and I want to work in accumulate experience.

Miss Li: What starting salary would you expect if you are employed.

Mr. Zhang: The salary is 6 000 RMB per month

李小姐：我们的招聘广告要求这个职位的应聘者应当具有相当好的英语水平，那么，你认为你的书面英语和口语能力如何呢？

张先生：我学英语已经10年了，口语相当好，能把自己的想法流利地表达出来。

李小姐：你获得过技术资格证书吗？

张先生：是的，我有计算机操作证书。

李小姐：那你能告诉我你有什么特殊技能吗？

张先生：我有计算机操作经验，熟悉微软 Windows、Word 和 Excel。

李小姐：你学过哪种计算机语言？

张先生：我学过C语言。

李小姐：你有没有程序设计方面的经验？

张先生：有。我曾经在英特尔公司担任过兼职程序设计员。

李小姐：你都学习了哪些课程？

张先生：我学的课程有计算机科学、计算机软件等。

李小姐：你和别人相处得怎样？

张先生：我和别人相处得很好。

李小姐：你为什么想为我们工作？

张先生：贵公司在电子领域是领头企业，我想在有丰富经验积累的公司工作。

李小姐：如果你被录用的话，你希望月薪是多少？

张先生：我希望月薪至少应该是

at least.

Miss Li: What about your marital status?

Mr. Zhang: I'm still single.

Miss Li: Do you receive any other companies's offers ?

Mr. Zhang: Yes. But I'll choose your company if your company accepts me.

Miss Li: I've looked through your application material, and I am quite satisfied with your qualification. Have you got anything to ask me ?

Mr. Zhang: Yes. Could you tell me something about the job?

Miss Li: Yes, of course. You would be responsible for design and development of produces.

Mr. Zhang: Oh, I think I can do the work well.

Miss Li: Then that's all for today. We'll inform you in a week. Thank you for your coming.

Mr.Zhang: It's my pleasure to attend the interview. Hope to see you again.

Miss Li: OK. Thank you very much. Goodbye.

Mr. Zhang: Goodbye.

6 000 元。

李小姐：你的婚姻情况是？

张先生：我还是单身。

李小姐：你有没有收到其他公司的邀请函？

张先生：有的。不过贵公司录取我的话我会选择贵公司。

李小姐：我看完了你的申请材料，对你的条件很满意，你有什么要问我的吗？

张先生：是的。你能否告诉我一些有关工作的事？

李小姐：好的。你将负责产品的设计与开发。

张先生：我想我一定能干好这份工作。

李小姐：那么今天的面试就到这里了。我们会在一周之内通知你结果。非常感谢你来参加公司的面试。

张先生：参加贵公司的面试我感到十分荣幸，希望能再次见到你。

李小姐：好的，非常感谢，再见。

张先生：再见。

Situation Dialogue

7.5 Reading and Compacting（对照阅读）

7.5.1 Resume（简历）

SHEN BIN
100 Zhongshan Road Shanghai　　Tel: 32114568
E-mail: shenbin@163.com
Objective: Sales Engineer

Education:

Beihang University, Beijing

Degree: BS,1999 Major: Materials Science

Beijing University of Iron and Steel, Beijing

Degree: MS, 2002 Major: Steel Making

Working Experiences:

2002-present Assistant Engineer, Shanghai Aeronautical Electrical Apparatus Plant

Personal Information: Date of Birth: Dec.2,1976; Male, single

Foreign Language: English, French, and German

personality: Happy disposition, outgoing, tactful

沈宾

上海中山路100号，电话：32114568

电子邮件：shenbin@163.com

求职目标：销售工程师

学历：

1999年理学学士 北京，北京航空航天大学材料科学专业

2002年理学硕士 北京，北京科技大学炼钢专业

工作经历：2002年至今 上海航空电器设备厂 助理工程师

个人情况：1976年12月2日出生，男，未婚

外语：英语、法语、德语

性格开朗、外向，为人机智

7.5.2 Notice（通知）

Meeting Notice

August 25

All personnel of the company are requested to meet in the First Conference Room at 9:00 a.m. on Monday, Sep.5. The subject of the meeting is to discuss the reform plan of the company.

Present's Office

会议通知

公司于 9 月 5 日星期一上午 9 时在第一会议室召开全体大会，会议的主题是讨论公司改革方案。要求全体员工出席。

总裁办公室

8 月 25 日

7.5.3 Visiting Card（名片）

Daqing Oilfield Company Ltd.

Zhang Ming

Director of General Office

Add: Beijing Road, Daqing, Heilongjiang,163453 P.R.China

Tel: 86-459-5951152

Fax Tel: 86-459-5951152

E-mail: Zhangming@petrochina.com.cn

中国石油大庆油田有限责任公司

张明

办公室主任

地址：中国黑龙江省大庆市北京路

邮编：163453

电话：86-459-5951152

传真：86-459-5951152

电子邮箱：Zhangming@petrochina.com.cn

7.5.4 Fax（传真）

场景一：（发传真）

Daqing Oilfield Company Ltd.

Tel: 5951152　　　Fax: 5951152

E-mail: botong@petrochina.com.cn

TO: Weiye Trading Co.　　　FROM: Wang Ming

ATTN: Mr. Jian Zhonglin　　　REF NO.: 001

TEL: 5951152　　　NO.OF PAGES: 1

FAX: 5951152　　　Date: 18 AUG,1996

RE: flight ticket

Your fax of 960815 was received.

We have purchased the flight ticket to Tokyo on CJ6320 at16:40 Nov.3 for Mr. Smith with RMB 4400. Mr. Han from our company will meet Mr. Smith and give the ticket to him at Beijing Airport.

Best regards.

Wang Ming

大庆油田有限责任公司

电话：5951152　　　传真：5951152

电子邮件：botong@petrochina.com.cn

发往：伟业贸易公司　　　发自：王明

收件人：简忠林先生　　　文件编号：001

电话号码：5951152　　　传真页数：1

传真号码：5951152　　　日期：1996 年 8 月 18 日

事由：

您 1996 年 8 月 15 日发来的传真已收到。

史密斯先生飞往东京的机票已购买，航班 CJ6320，日期 11 月 3 日，16:40 起飞，票价为人民币 4 400 元。我公司的韩先生将在北京机场与史密斯先生会面并将机票转交给他。

最诚挚的问候。

王明

场景二：（回复传真）

To meet Mr. Wonkyu Lee, CEO of KNPC

DQCL request the pleasure of the company of Mr. Harris Smith at a Tea Party on Friday, September 5, from 3:00 p.m.to 5:00 p.m. in Room 207, Daqing International Hotel.

哈里斯·史密斯先生：

为欢迎韩国石油天然气总公司首席执行官李龙渊先生，兹定于 9 月 5 日下午 3:00 ~ 5:00 在大庆国际饭店 207 室举行茶话会。敬请光临。

中国石油大庆油田有限责任公司

回复样本：

Mr. Harris Smith

Accepts with pleasure the kind invitation of DQCL at the Tea Party to meet Mr. Wonkyu Lee, CEO of KNPC on Friday, September 5, from 3:00 p.m. to 5:00 p.m. in Room 207, Daqing International Hotel.

大庆油田有限责任公司：

我很高兴接受贵公司的邀请，参加 9 月 5 日下午 3:00 ~ 5:00 在大庆国际饭店 207 室举行的欢迎韩国石油天然气总公司首席执行官李龙渊先生的茶话会。

哈里斯·史密斯

7.6 Professional Terms（专业术语）

1. Nationality：民族，国籍
2. Home code：住宅电话
3. Business phone：办公电话
4. Marital status：婚姻状况
5. Educational history：学历
6. Major：专业
7. Social practice：社会实践
8. Summer jobs：暑期工作
9. Typos：打字稿
10. Computer Data Input Operator：计算机资料输入员
11. Computer Engineer：计算机工程师
12. Computer System Manager：计算机系统部经理
13. Data Processor：资料员
14. Chief Engineer：总工程师
15. Hardware Engineer：硬件工程师
16. Secretary：秘书
17. System Engineer：系统操作员
18. Software Engineer：软件工程师
19. Typist：打字员
20. Committee of Science and Technology：计算机集团公司
21. Computer Technology Service Corporation：计算机技术服务公司
22. Commercial School：商业学院
23. Foreign Language School：外语学院
24. University of Science：理科大学
25. Department of Computer Science：计算机科学系
26. Department of Electronics：电子系

Exercises（练习）

1．Complete the following dialogues.（完成下列对话）

（1）A：When will you graduate from college?

B: ________________________________

（2）A:________________________________

B: The subject I like best is computer science.

（3）A:________________________________

B: I got an average of 85 points.

（4）A:________________________________

B: Yes, I was the class commissary in charge of recreational activities.

（5）A: Were you involved in any club activities at your school?

B:________________________________

2．Answer the following questions according to your personal information.（根据个人信息回答下列问题）

（1）What's your name, please?

（2）Where are you from?

（3）How old are you?

（4）What school are you attending?

（5）What subjects have you learnt at college?

（6）What subjects do you like best?

（7）Did you get any honors or rewards at your school?

（8）Have you got an excellent record in English?

3．Please write a resume according to your personal information.（根据个人信息写简历）

个人简介

姓　名：×××　　　　性　　别：女

民　族：汉族　　　　政治面貌：团员

学　历：本科　　　　专　　业：计算机

联系电话：139512888888

联系地址：四川省广元市

邮　　编：628017

电子邮箱：xyz@163.com

学　　历：

毕业学院：四川信息职业技术学院　2009.9～2012.6　计算机专业

专业特长：

1．计算机工程师职称，熟悉计算机软件编程

2．英语4级，熟悉各种英文函书写

3. 精通计算机基础知识，能熟练使用 SQL Server、C 语言、Office 办公软件

4. 头脑灵活，有责任心，有较强的团队精神

经　　历：

2012.6 ~ 至今　ABC 集团公司 计算机管理员

2009.9 ~ 2012.6　四川信息职业技术学院

性　　格：

开朗、谦虚、自信

附文 7: Reading Material（阅读材料）

Reasons For People Don't Get Hired
（人们不被聘用的原因）

1. Poor personal appearance 糟糕的外表
2. Over-aggressiveness 过于放肆
3. Lack of interest and enthusiasm 对工作缺乏兴趣和激情

4. Lack of planning for career—no purpose or goal 对工作缺乏计划，无目标
5. Nervousness，lack of confidence and poise 紧张，缺乏自信，不够沉着
6. Lack of tact and courtesy 缺乏机智和礼貌
7. Lack of maturity 不成熟
8. No eye contact with the interviewer 与面试人员缺乏眼神交流
9. No sense of humor 缺乏幽默感
10. Arriving late for the interview 面试迟到
11. Vague responses given to questions 回答问题模糊不清
12. Negative attitude about past employers 批评以前的老板
13. Failure to express appreciation for interviewer's time 不感激所给的面试机会
14. Inability to express information clearly 语言表达不清
15. Overemphasis on money 过分强调薪酬

Practical Examples of Resumes

Resume of Xiaoxiao

Dept. of Computer Engineering
Sichuan College of Information Technology
Guangyuan City, Sichuan Province
Telephone:139××××8852
Career Objective: Software Engineer

Qualifications:

1) MS degree in Computer Science
2) Two years of part-time work experience in IT industry
3) Solid knowledge of programming and computer languages
4) Self-motivated and capable of working independently as well as in a team

Education:

Master of Science in Computer Science Expected March 2011 Sichuan Information Technology

College:

Major Courses included: Computer Science, C,C++,Operating Systems, Systems Management

Skills:

Familiar with computer programming and personnel administration softwares
Good command of English

English Proficiency:

Fluent in reading and speaking College English Test Band Six

Interests:

Enjoy playing tennis, reading and listening to classical music

Personal Data:

Age: 23 Sex: Male Health: Excellent Height: 180cm Weight: 70kg